AI DRIVEN

Wolf Ruzicka
Victor Shilo

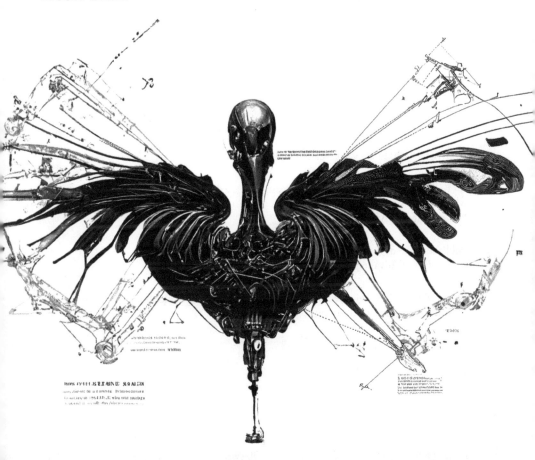

AI DRIVEN

Staying Alive in the Age
of Digital Darwinism

Advantage | Books

Published by Advantage Books, Charleston, South Carolina.
An imprint of Advantage Media.

ADVANTAGE is a registered trademark, and the Advantage colophon is a trademark of Advantage Media Group, Inc.

Printed in the United States of America.

10 9 8 7 6 5 4 3 2 1

ISBN: 978-1-64225-378-8 (Hardcover)
ISBN: 978-1-64225-462-4 (eBook)

Library of Congress Control Number: 2024904907

Cover design by EastBanc Technologies.
Layout design by Matthew Morse.

This publication is designed to provide accurate and authoritative information in regard to the subject matter covered. It is sold with the understanding that the publisher is not engaged in rendering legal, accounting, or other professional services. If legal advice or other expert assistance is required, the services of a competent professional person should be sought.

Advantage Books is an imprint of Advantage Media Group. Advantage Media helps busy entrepreneurs, CEOs, and leaders write and publish a book to grow their business and become the authority in their field. Advantage authors comprise an exclusive community of industry professionals, idea-makers, and thought leaders. For more information go to **advantagemedia.com**.

CONTENTS

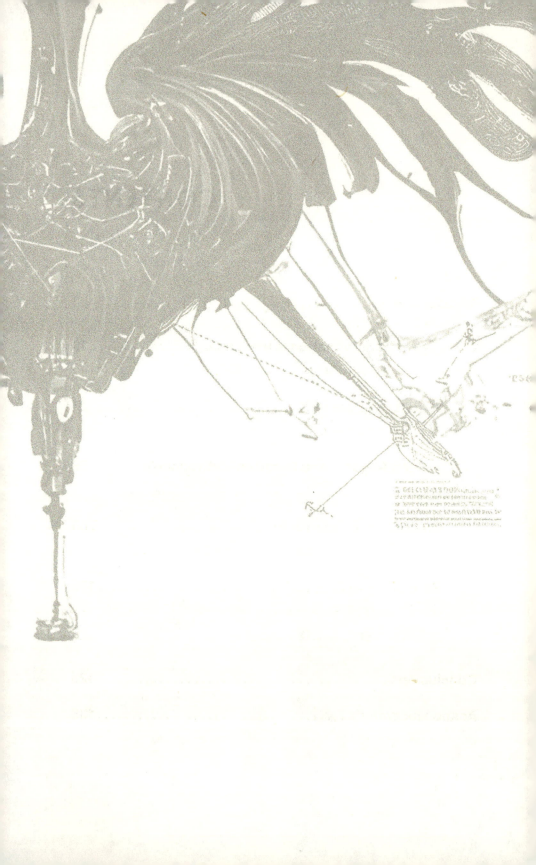

INTRODUCTION

A Matter of Life or Death

Imagine for a moment that you're a small mammal scampering around the brush on a fine sunny day sixty-six million years ago. Life is hard. You constantly scavenge for food and hide and run from the dinosaurs who dominate the food chain. That is, until one day something bright streaks across the sky.

The asteroid impact was cataclysmic, and the subsequent climate changes created the Cretaceous mass extinction event that took out an estimated 78 percent of the world's species. Among the many victims was the T-Rex. Those ferocious Velociraptors. The giant Diplodocus that nearly crushed you a few days ago. Only those that found ways to adapt to the radical changes of the new post-meteor-strike world—such as birds, crocodiles, and turtles—lived to see their grandchildren and great-grandchildren. While the fearsome giant reptiles were gone, more adaptable species were about to enter their heyday.

This is basic Darwinism—survival of the fittest—or, in this case, the most adaptable, when disaster strikes. Those who adjust to environmental changes survive. Why do we begin this book by talking about the extinction of the dinosaurs—and the rise of mammals?

What does it have to do with AI-driven Digital Transformation in the twenty-first century?

Well, everything.

We're not exaggerating when we say we believe that AI-driven Digital Transformation is going to mean life or death for companies. Companies that adapt will be able to survive business fragility and market turbulence, hallmarks of what we call "the Age of Digital Darwinism."

At the most fundamental level, Digital Transformation is directly tied to Digital Darwinism—the companies that use technological adaptations to face a changing market will survive, while the others will go the way of T-Rex and co.

Embarking on an AI-driven Digital Transformation journey may seem overwhelming, daunting even, but you are not going to lose your position over starting this journey. However, if you don't have the stamina to go through with it or if someone else beats you to being the disruptor in your industry, then you could or your business could go bankrupt. If you follow the concepts laid out in the next five chapters, then you will understand the various steps of the journey we fully anticipate are inevitable for a growing company and be able to lead the AI-driven Digital Transformation that will enable your company to not only survive but thrive.

Technology and engineering services our team members have provided power the successful AI-driven Digital Transformations of companies across the globe. Most of our work takes place behind the scenes. In this book, we will share some of our successes and lay out practical examples of AI-driven Digital Transformation that you and your teams can use for inspiration.

You don't need to be an AI expert, and we don't want to turn you into a data scientist. Our goal is to educate and equip you to better understand the technical side of your business so you can effectively contribute to your company's AI-fueled Digital Transformation.

You may be afraid that you are too late to the party, that you're too far behind. You're not. Our best advice is to jump in today—or appreciate the high risk you *will* be left behind.

So buckle up and read on. When you finish this book, you will have a clear understanding of:

- The process of Digital Transformation
- The basics of AI and the essential role of structured data
- Multiple best practices to quickly adopt AI
- Unstructured data as a reliable source for AI
- Accelerating creative business processes using generative AI
- How to prosper in the Age of Digital Darwinism

From Point A to Point AI

The foundation of *AI Driven* is understanding the need to adopt AI as a foundational element of Digital Transformation now and use its prediction power to adapt how your company operates across teams and maximize your business value—shareholders will be delighted—and have more informed decision-making across all your processes.

Today's business leaders—from the CEO to the CMO and head of people operations—must play a role in adopting AI to ensure the survival of their company. By sharing real-life examples of the simple steps of how businesses undertake Digital Transformations while adopting and implementing AI to increase ROI and identify new revenue sources, we will show you how AI is already being used—and how you can prepare yourself to leverage AI for maximum impact, in an understandable format.

In writing this book, we aim to close the knowledge gap for CEOs and senior business leaders to foster effective cross-functional AI strategy and execution within their teams. Companies that put these practical approaches to use will mitigate the traditional data and AI silos separating business and technology executives, and their related metrics, to not only survive but thrive in today's digital business environment.

These are not theoretical concepts. We use real examples of real companies we've worked with so that you can apply these key learnings to your business. Using five case studies, we will lay out the leading principles, practical implications, and business benefits of AI-driven Digital Transformation. These companies' experiences exemplify key concepts and imperative actions that will help your business take the necessary steps to remain relevant, competitive, and future-proof.

Equipped with this new knowledge, you will be able to attend executive leadership meetings and intelligently navigate or even direct the conversation, not simply nod your head. We want you to go from having superficial knowledge gleaned from a business article to being able to conduct an informed conversation, whether the topic is data science, machine learning, enterprise software development, cloud infrastructure and services, the importance of leading architecture, or the unifying role that AI has in relation to all these disciplines.

Processes, Not Humans

AI-driven Digital Transformation can be perceived as a broad and complex set of topics. For the purposes of this book, we carefully selected five clients whose journeys best illustrate the key principles of successful Digital Transformations. To be able to give our readers an unvarnished look into the dos and don'ts, successes and failures, and their lessons learned along the way, we withheld the client identities. The case studies not only illustrate how others are benefiting from AI already but also help you assess whether your business is ready to implement AI as part of its own Digital Transformation.

- *Case Study One—Digital Transformation Is Inescapable*
  Client: Global News Outlet
- *Case Study Two—Structured Data for AI*
  Client: Public Transit Agency
- *Case Study Three—AI-First Mindset and DataOps Methodology*
  Client: Multinational Financial Company
- *Case Study Four—Unstructured Data for AI*
  Client: Innovative AI Patent Clerk
- *Case Study Five—Generative AI*
  Client: Revolutionary AI-Driven Curriculum Developer

As we get into our case studies in the following chapters, you'll experience the tangible benefits AI-driven Digital Transformations offer any organization, not just the five we picked as our examples. For decades, some form of AI has been present, but only recently has it captured the public's mindshare and inserted itself more prominently into everyday tasks. AI is not only here to stay, but more and more AI and advanced AI are rapidly emerging and ready for the public's, corporations', and governments' adoption—if they are ready and start preparing today. Our objective is to help you see how your systems, your solutions, and your business can be prepared for this critical next step.

There is a fear sometimes that increased automation will make humans redundant or that AI will overpower mankind and become our overlords like the machines in *The Matrix, The Terminator*, or *2001: A Space Odyssey*. These suggestions do not concern us. Those dystopian movies—even though they explore some of the most mesmerizing possibilities of AI—are pure entertainment. In reality, every AI needs a human master to learn from—it has an on and off switch and is designed to serve somebody. There is no indication yet that machines will become self-aware and detach themselves from human control in the foreseeable future—if ever. The human master will remain the master, even if the AI becomes better than a human at certain tasks, such as playing Go, ordering supply chain replenishments, or assessing other vast and complex decision options in order to allow an executive to pick the one with the highest probability of success.

AI is not a solution to everything and is not required everywhere. Throughout our case studies you will see a trend of focusing our work on what is needed and immediately achievable to start and growing from that point. Many manual tasks do not require an AI to be automated. The house robot of the future such as Rosie, the Jetsons' classic maid robot, could be designed to automate many common tasks such as laundry or cleaning the dishes. Nonetheless, it was more expedient to have Rosie use a washing machine or dishwasher to accomplish the task. Of course, you could teach an AI to do your dishes or wash your underwear. But why? It is a waste of your creativity, time, energy, and resources. Any solution to any problem or automation of a task should be as simple, efficient, and cheap as possible as measured per dollar, watt, or minute. Solutions should be economically viable, easy to accomplish and maintain, allowing AI to solve much, much bigger problems with the highest possible impact, not just how to better wash a spoon.

This will be equally true with how AI assists industries like banking or transportation. It will not kill their systems, but adapt them to be more efficient and powerful, which ultimately helps the business. What will harm your business is if someone else beats you to the punch.

In the end, AI exists to serve your business and you, whether by providing you with better media content or by automating some of the mundane, repetitive processes you don't find value-creating to do yourself. In a business context, there can be some trepidation over the implications of using AIs. Some worry or enthuse over whether AI will eliminate jobs.

There are portions of your company—processes really, not humans—that should be replaced or upgraded by AI over time. They lay the foundation you can build upon to make the entire company more creative, more efficient, nimbler, and easier to manage. AI is not replacing humans, but rather helping us reallocate our talents and skills to more creative, stimulating, and value-adding areas.

Thriving in the Age of Digital Darwinism

While AI has to be born inside the core of your company, it also has to move to the *edge* of your business aggressively. It has to come from the core of what you do, the foundation of your work. At the same time, AI has to move to the edge where you are interacting with your customers, your partners, your employees, where the digital and analog worlds meet. Here it remains invisible, working quietly but efficiently. This is what we call "AI on the edge."

We know this sounds risky. And it *can* be risky, because not every company is ready to dive into AI right away. Over time, we developed and battle-tested effective methodologies for de-risking this process. The AI readiness assessment and the MVP (Minimum Viable Prediction) are two methodologies that we follow to minimize the risk and constantly ensure progress toward successfully adopting AI.

Additionally, much of the risk can be mitigated through good DataOps, which is one of the key pieces we examine in the AI readiness assessment. Many organizations fail, because they rush to gather and analyze their mountains of data without a plan when they would be better off narrowing their focus to find the relevant data "breadcrumbs"—the right minimal data that will direct the best path toward developing an MVP, something that can be done in months rather than years.

Your primary safety is in controlling and accelerating your speed of change. The risk of extinction is dramatically higher if you don't evolve.

This brings us to one of our favorite books, *The Black Swan* by Nassim Taleb, in which the author describes "The Black Swan" as an improbable but incredibly impactful event that, when it happens, destroys things—like the dinosaur-killing asteroid. Digital Black Swans, while equally rare, have the ability to immediately highlight the vulnerabilities in a system, often catastrophically (i.e., Y2K). Infrastructure requires constant focus on Digital Transformation to moderate the risk of such events. In technology, disruption is always present in varying scale. Months after we started writing this book, ChatGPT emerged as only the latest and surely not the last Digital Black Swan that only few seem to have seen coming. Launched in November 2022, ChatGPT's generative AI tool instantly hit the public consciousness with dizzying effect, disrupting the most prominent, seemingly all-conquering search engine in the world, which seemed untouchable until the ChatGPT swan descended. ChatGPT's disruptive presence won't end there. The tool has penetrated research, higher learning, creative arts, and, yes, book publishing with asteroid-like force. Prompt engineering has emerged as an already overhyped discipline. As a result, those sectors and individuals impacted have had to up their game to detect and prevent plagiarism and find ways to coexist with an all-conquering AI.

Digital Transformation: Survive and Thrive in an Era of Mass Extinction by Tom Siebel is another book that influenced our thinking early on, where the author examines some of the emerging technologies that are disrupting industries and how they can be harnessed to transform organizations into digital enterprises that are constantly prepared for the next unpredictable, highly disruptive event when—not if—it explodes on the scene.

These two concepts return us to the topic of Digital Darwinism in business today. Technology innovations are the asteroids that kill legacy companies that have not prepared to continuously adapt. Some companies have followed the processes we lay out and therefore are in a perfect position to adapt, transform, and thus not only survive but continue to thrive by confidently taking advantage of such disruptions instead of fearing their arrival. Others fail to act—and risk extinction.

Businesses can be fragile, and markets are unpredictable. Just look at the seemingly-out-of-nowhere collapse of Silicon Valley Bank in 2023. In hindsight we know what happened, but nobody predicted it, and only with brute force at lightning speed—and asteroid-size resources—did the impact get minimized. With AI, you have the opportunity to better predict your next move and turn yourself into what Taleb calls an "anti-fragile organization." These are the mindsets and tasks of being set up in a way that you *benefit* from these Black Swan events rather than being killed by them.

In the Age of Digital Darwinism, an infinite Digital AI Transformation—it never stops, because technological innovations never stop—is the adaptive mechanism that allows you to survive and thrive beyond the cataclysmic event, whatever it is. The investment may have the ultimate ROI: your and your organization's dominance. The frightening alternative is to vanish into oblivion and reappear as an exhibit like T-Rex Sue in Chicago's Field Museum of Natural History, where the dinosaurs that survived—like birds, crocodiles, or turtles— and the mammals that thrived after your extinction will marvel at your once mighty history.

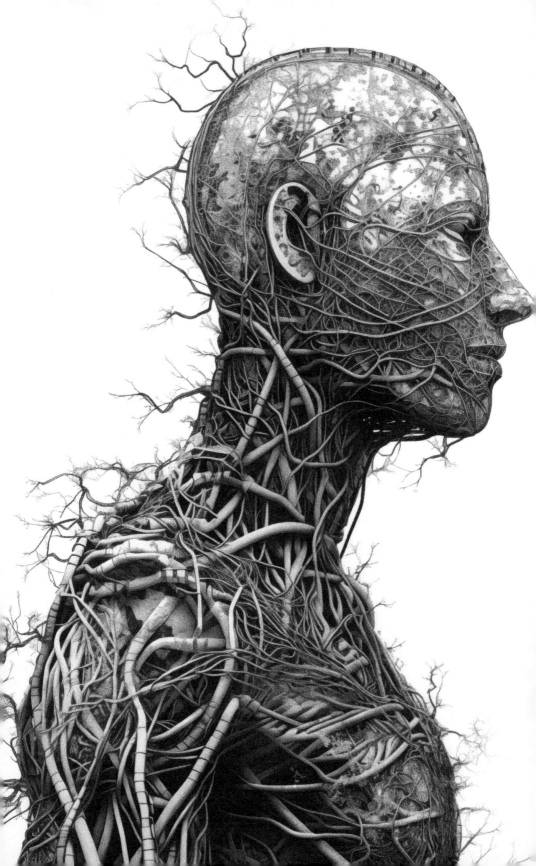

Digital Transformation Is Inescapable

Case Study One: Global News Outlet

Imagine a traditional print media company under threat from constant attacks by native digital news channels. The effect of Digital Darwinism has been brutal on journalism, which has been hit with an earth-shattering effect. From local dailies to storied, trusted representatives of the Fourth Estate, newspapers have had to fend off competition from nimble internet-based start-ups and new digital media seeking to reinvent news reporting and delivery and compete with free news aggregation websites—and, of course, new, easy-to-digest media formats. Attention spans have shortened. Countless papers have downsized or perished. Some have reinvented themselves and survived. A few have even thrived while their competition withered away. The dinosaurs are gone—only those that could adapt to the new world are still alive.

In this chapter we introduce a necessary understanding of Digital Transformation and prepare our reader for the entry into AI topics, highlighting how technical evolution is possible and, in many cases, necessary while maintaining and protecting an active system and business. Digital Transformation in itself is not a new or revolutionary concept, but its complexity and necessity is accelerating with the advent of AI.

Our first case study concerns our work with one of the world's top news organizations. Facing the very real threat of decline like so many others, this organization accepted the urgent need to reinvent itself. We helped it successfully transform itself from a traditional news organization into a software company—a fundamental change of identity that not only enabled it to fend off extinction but also to become a pioneer in how news is delivered. Here is a real-life scenario of a global news outlet moving from outdated legacy systems to an evolutionary state where they are now considered the leader in their space.

Rather than seeing the client as a news company, we saw it as a software company where its journalists were the customers of the platform, providing them the necessary tools to share their content faster and more seamlessly across much wider channels with extreme personalization. That way, they were able to provide the audience with the most relevant content, delivered in the format that suited each user the best. While AI was not yet part of the initial effort, the case study illustrates that Digital Transformation is the necessary step to becoming AI-ready.

In this chapter, we'll unpack its transformation, which addresses how:

- Digital Transformation enabled a corporate identity shift

- Legacy systems were replaced without interrupting daily business
- The news organization created an alternative software revenue stream

Becoming a Software Company

Traditionally, software advancement has been driven by little start-ups, such as fintechs disrupting the big banks. But with Digital Transformation, the tables have turned. You can be the disruptor in your own industry by developing your own software tools and leveraging innovative cloud services. You have the power to do it. Instead of waiting for the next big thing, you could *be* the next big thing.

Through Digital Transformation, you translate your IP to software. Sometimes, you can even monetize that IP, which means you can license that software to others, even former competitors. If you can pull it off, this shift in acting as a software company first means you would actually benefit from the strength of your own competitors because the better they are at adopting your software, the more revenue you will make, the higher the valuation of your company will be.

In other words, if you are one of the first in your industry to truly take Digital Transformation seriously, not just give it lip service, there is a possibility that you can create real growth and generate a new recurring revenue stream. If your competitors beat you to it, you could be the one licensing their software. Businesses that fail to make this shift to viewing themselves as a software company first may not die, but they could sacrifice their competitive edge and miss out on an incredible business opportunity.

The term *Digital Transformation* is thrown around a lot in tech and business circles—but what it actually refers to can be confusing. Does it mean adopting tech to improve the customer experience? Replacing humans with software? Automating business processes? Implementing software solutions designed to eliminate systematic inefficiencies?

Yes—all of those can be *aspects* of Digital Transformation. Rather than seeing Digital Transformation as a "project," it creates a shift in your very business identity at the core of what you do.

Digital Transformation is the conversion of your company to a software product company. In other words, if you are a bank, you turn yourself from "We are a bank" to "We are a software product company that provides financial services."

As a software product company, you constantly automate necessary activities based on your priorities and by taking an optimization-first approach. You look at how everything *should* work, contrast it with how it currently works, and create an aggressive road map to fill the gaps. It's about finding new ways to run things better tomorrow than you did the day before. Every day.

The Digital Transformation of this client went beyond just "updating" its systems as usual and instead created a total shift in its identity. It successfully went from "We are a news company" to "We are a technology company that happens to have a newsroom full of reporters." This reinvention meant it not only dramatically added subscribers and industry prestige but also generated an alternative revenue stream that did not exist prior to its transformation. How? By developing its own software, which it then licensed to others.

The Core

Before its Digital Transformation, this client was a very traditional news organization centered around its reporters, newspaper delivery, and advertising. Everything in the digital space was an afterthought, but not for lack of spending large budgets on IT.

The underlying motivation for the client to undertake this journey was to prevent a slow death. While the organization had an illustrious history, and its journalistic standards were revered in the field and by its readers, the news *delivery* method remained stuck in the past. The outlet was facing a very real threat of extinction. Its steady decline was exacerbated by digital attackers coming out of the woodwork. New outlets and platforms like Google News, Vox, Facebook News, or Snapchat Stories were taking hold of the news world. The revolution in the media industry forced established players to rapidly adapt or slowly and painfully diminish. Luckily for our client, a few members of its leadership team had a resolute will to evolve.

Digital Transformations should not start only after you completed your detailed plan step-by-step; that may be too late. It also shouldn't happen everywhere all at once. Your organization will be paralyzed by confusion. It starts in the technical core of what constitutes your organization's unique identity in parallel with business planning and evolution. If you are a bank, those are your banking core systems. If you are an online retailer, those are your e-commerce systems.

The core of our customer's business was a content management system (CMS), licensed from someone else. It pulled content from the reporters to create a PDF that could be sent to city-block-sized printing machines. The news portal and mobile apps were an afterthought, and their content a derivative of the main content production process.

17

The outlet's website, mobile apps, and print media were connected to this core CMS—it was like the "one ring to rule them all." As the news outlet steadily kept losing subscribers and revenue, it became clear to us that this core absolutely needed to be replaced, either through a newly designed replacement or as a last resort, by another vendor's core—we had to bring it to Mount Doom and recast it.

We advised against a traditional "lift and shift" approach to a CMS from another vendor, which can appear easier to scope but can take a long time, risking that the replacement system might already be outdated. Appreciating the speed of transformation is increasingly relevant as AI seeps into the marketplace, requiring the approach we recommended. Some people embrace the cautious approach, as it provides near-term job security even if it dangerously impairs the business. Multiyear, multimillion-dollar lift and shift efforts at the core of a business to us feel like painkillers, but not like a cure to the company's underlying disease.

At the news outlet, it was essential to identify which capabilities the CMS was missing and which of its existing capabilities should be more efficient in the digital realm. Those gaps informed a road map of small, inexpensive, lightweight modules that needed to be created within months and with minimal budget impact. The moment one such module was identified, the engineering would start while other modules were identified in parallel. Ultimately, the modules would surround the old CMS and strangle it to death. Small investment with quick turnaround meant that if we got something wrong, we would learn from it quickly and prevent the same mistake in the future.

Within a few months, the first module was in production, a portion of the old CMS was already bypassed, and the transformation into a software product company had started.

Grow the New, Strangle the Old

The rapport and trust we had developed over several years of partnership allowed us to work on the client's infrastructure and the very core of their business. When we got this type of intimate access to their backend API development and looked at their systems, it became obvious the technical options were limited. Every day the system was in maintenance for an hour and new content could not be deployed, negatively affecting the subscribers' experience. Breaking news could not be broken during this period. It was clear to us that the company needed to fully transform its systems to be able to easily add the most cutting-edge technologies on the market and be accessible to its consumers every hour of every day of the year.

We methodically split the client's big, siloed CMS into logical subsystems, which would be created with only the most necessary dependencies on older systems, so they would evolve independently at their own pace, and be easily interconnected with each other. Each subsystem was optimized to handle a specific task the old CMS either could not perform or performed in a suboptimal way.

While there are several other technical approaches to replace clunky old legacy systems than the traditional lift and shift approach, this approach is one of our favorites where applicable. As violent as it may sound, we call it the "Strangler Pattern," an analogy to visualize how to replace legacy systems with new, agile modular systems that enable a company to transform digitally without interrupting day-to-day operations and processes.

Strangler figs, whose aerial roots wrap around their host trees, form a latticework that eventually kills the hosts, leaving the thriving figs encircling the dead, often hollow tree trunks. Think of clunky legacy systems like this customer's core CMS like big, rotting old trees, which need to be cut down to make room for new growth.

19

Following such a "Strangler Pattern," we left the media company's existing CMS in place like an old tree and built the various modules around it like a fig tree's roots. Some of the modules were designed to replace tasks the old CMS performed. Others performed new tasks it could not accomplish, such as omni-channel distribution of content across all social media platforms, gathering real-time reader statistics, or addressing video delivery issues with code that could easily be plugged into articles and other content.

We launched these new subsystems into production in parallel to the client's existing CMS, and they soon became the new standard, gradually replacing older technologies and processes. These smaller modules made traditional features of the old CMS obsolete and allowed us to safely decommission them. The former one-size-fits-all approach for their content across all platforms was replaced with this new concept of optimizing content for its destination. We kept replicating the same approach with other pieces of their business, such as images, article production, live news coverage, irregular news events requiring specific media, and new formats of illustrating complex topics, optimizing the entire portfolio of processes by making the ensemble smarter and nimbler to orchestrate.

In the end, a few dozen relatively small, lightweight modules that were easier to maintain and enhance independently made up the core of the news organization without causing downtime or straining the client's budget. Its technical options were now virtually unlimited, and its continuing transformation had unlocked many opportunities to experiment with technical and business model innovations.

Optimizing the Delivery

One of the important aspects of the project was improving the client's digital response time. Before the transformation of its CMS, the client's website took a few seconds to fully load. No matter how good your content is, this is not an ideal user experience. When a website—especially a news site—has an extremely fast load time, it becomes a technical advantage that others benchmark themselves against.

Improving the load time was key to helping our client gain organic growth in its audience. Its content had not changed—it had the same writers, same editors—only the speed was faster. This speed directly connected into the client's organic growth on sites like Google and through social media where the algorithms are designed to prioritize sites that load faster. Speed gave our client's content preference on other platforms. The result: more audience engagement, more shares on social media, and ultimately more paying subscribers.

Having successfully optimized our client's content delivery, the next question was how to maximize the speed and reliability of deploying changes to the system. The solution was to remove the human factor and *automate* these processes. This highlights how attention to Digital Transformation can increase opportunities to introduce modern AI.

This happened gradually. Throughout this client's Digital Transformation, the deployment of changes in the system became more and more automated. Prior to the Digital Transformation, a team of humans deployed any changes in the system. Once we implemented a technical pipeline for deployment, we could remove the human factor from the production line, and then whatever changes needed to be made were rapidly pushed out within minutes. Imagine being able to update any and all of your enterprise systems within minutes every few minutes without turning them off. This both improved reliability and freed up human capital to focus on other needs.

Today, our client's system deploys everything in a repeatable and clear way that gives its stakeholders a sense of safety and reliability. There is safety because there is no longer dependency on a singular individual to make sure it's deployed correctly. And there is reliability because you know things will be deployed in the same way as the last time—you don't have to come up with a new deployment plan or resources every time.

Designed this way, automation of the deployment process adds order. And with order, you add predictability and, therefore, less risk. In some cases, a human is still involved, but you should automate as much as possible, because the human factor is where you are most likely to have an error and a bottleneck.

On another level, the transformation helped our client grow its capacity as a team. In the beginning, its staff had no software engineers or machine learning (ML) engineers. There was an IT department buying, installing, and maintaining products from other vendors, but that was all. Software and ML engineers are very creative people—they seek out problems and solve them with their creativity—not the run-of-the-mill IT needs like the notification you need to update your operating system.

As further evidence of our client's identity shift, it began embedding its engineers in the newsroom. At a newsroom desk of six, five might be reporters and the sixth a software engineer, so they can learn from each other. This was full integration of technology and business teams, which is exactly what you need in order to turn yourself into a software company.

As we predicted, during one of our catch-up meetings with the CTO, he shared with us, "Competitors are calling me and saying,

'Your content is the fastest-loading in the industry and gets shared most on social. We've been trying to replicate what you did and can't figure it out. Can we just license your software?'"

There were several layers of organic growth resulting from this transformation. Throughout this process, the outlet's digital subscriber numbers started accelerating, at some point even eclipsing its competitors' in the digital world. But in addition, the solutions we jointly implemented led to the creation of alternative revenue streams that scaled to many hundreds of millions of dollars per year by successfully monetizing and licensing the new publishing software platform. The struggling news outlet that had lost its shine had now become a profitable, growing king of digital news.

To the Cloud

When we started this project, we knew we had to find the best way to future-proof the client's systems so that it could gain a significant and lasting advantage over other media companies. We suggested an approach that has been a staple of every modernization: moving to the cloud, which would provide a more robust infrastructure, allow us to add innovative web services, and give the ability to scale up and down when needed. For the client, this approach was an entirely new one. They had everything in their data centers, so they considered moving to the cloud a new, and therefore somewhat risky, concept. But *not* moving into the cloud meant recreating some or all the tools and services that most modern cloud providers already provide off the shelf—time and money that would be better spent elsewhere.

Everything we were building at the time—mobile app services, modules—was deployed in the cloud. If we needed to tweak a system, we did not have to rebuild the entire infrastructure—we could focus on only optimizing that one piece. Take videos, for example. These had to be converted to different platforms at the time—iPad, iPhone, website, social networks—and doing so required a lot of CPU or GPU power, which became very costly. The flexibility of the cloud alleviated this as we only paid for the power we needed to complete these processes. If some of the videos were not as time-sensitive, the system was designed to convert them at night when the cost was cheaper.

When people think of the cloud, they often think of it in terms of hosting. The cloud is not just somebody else's computer. The purpose was not so much to host the client's content but to transform its systems through cloud services with the power to make them faster, more intelligent, and more efficient—not just during the course of the implementation, but forever, because due to economies of scale you take advantage of everyone else's innovations.

Furthermore, the various cloud services became the tool with which we could peel away the old legacy systems and no longer have to constantly "lift and shift," because the optimization is automatic, and the communication between the individual functions is more seamless.

On a more general level, a key benefit of a modern cloud is that it provides a widely applicable framework, which the user tailors to their specific requirements and objectives. Using feedback loops and generated data, you can learn how to customize the cloud service to your needs so you build your own "secret sauce" on top of it.

This is where your own software engineering team comes in to learn from others what needs to be tailored to you, after which they can build those layers themselves. They are not starting from scratch—they can better apply their creative problem-solving to optimize the service to the business needs, compounding on the benefits of Digital Transformation already achieved.

Choosing the cloud for this project posed no risk and proved to be exactly the right approach. The cloud services allowed us to build a true zero-downtime environment, which was essential in a twenty-four-hour news cycle. The system was built to last and will hold up for years to come. In short, taking a calculated risk and making every Digital Transformation a *Cloud* Transformation creates new, agile, and *future-proof* systems, ready for AI.

Digital Destiny

Over time, every business develops its own identity and intellectual property. Something unique that makes you, *you*. Your secret sauce. To survive and really thrive in the digital world, your objective should be to translate this "secret sauce" into your own software. The key to successfully owning your digital destiny—really your entire destiny—is to start owning your existing systems' core. If you license it from someone else, they own your digital destiny. Over a relatively short period of time, our media client embraced that evolution and got recognition for it! Its secret sauce was the creation and management of its reporters' content, and they took ownership of the core system responsible for it, rather than licensing it from someone else.

To turn the tables on existing competitors in your industry segment and to prevent digital natives from disrupting your business, you need to exhibit an unwavering will to survive on an executive level and be ready to start a journey of relentless small experiments. It starts with a commitment from the top and is driven forward by determination to never give up once you start. The daring Big Bang approaches that CEOs so often like to announce are where the highest risks lie. Just as any capital-deprived start-up has shown many of us, the power of change lies in the small, relentless iterations that spread across any given (legacy) technology stack, an existing team that may be hesitant to change, and entire industries that seemed unassailable until they weren't. We iteratively created small, lightly connected software modules within the core of the media company, but not inside of the core CMS system itself, so the company could start owning its digital destiny.

When transforming a big system in small steps, the key metric to keep an eye on is the rate of change after each iteration, which can have a compounding effect when managed well. It also reduces risk while maximizing the probability of succeeding. Imagine you manage to positively impact one of your KPIs by just a few percentage points over the period of a two-week sprint cycle by creating or adapting a small, lightweight module within your organization's core, but outside of the monolithic legacy system that manages your organization's key data and processes. This gain can now potentially be compounded at every engineering iteration. Your job now is to keep an eye on the rate of change from iteration to iteration. Keep it constant and you create a hyperbolic growth curve in the improvement of this particular KPI. Manage to increase the rate of change consistently, and your KPI will increase hyperbolically more quickly.

To further ensure an approach that manages your enterprise risk, new systems should be launched in parallel with existing systems. Just like with our news outlet client, these agile systems will strangle the legacy monoliths to death over time as the new software becomes more and more functional. You won't have to make that one risky, make-or-break decision such as "lift and shift." Rather, you make a series of relatively small prioritization decisions during each sprint cycle, that is, engineering interval. Another risk-mitigating factor—and, in fact, a factor for higher probability of success—is to automate the deployment of changes gradually.

Another key to successful Digital Transformation is to prioritize your efforts. Don't waste your and your engineers' creativity and brain power on recreating the tools and services that the modern cloud providers have already prepared. Instead, use your brain power and time to identify and transform your unique IP into software that you own and can perhaps monetize on top of that foundation. A range of cloud solutions provide this technical foundation for you to build upon.

For the storied, once stodgy media company, this successful Digital Transformation resulted in a meaningful alternative revenue stream. Due to our collaborative approach, we were able to transfer responsibility of the new system and its evolution to our client with ease with minimal need for support. With our help, they expanded their engineering team to truly own their digital destiny going forward and felt the pride of having become a software company with a newsroom full of reporters.

Executive Summary

I'm a CEO of a large organization. I'm reading the book called *AI Driven*. I need a very short explanation (three to four sentences) of why what is described in this chapter is important for AI.

This chapter underlines the critical importance of Digital Transformation for organizations facing disruption from digital competitors. It emphasizes that embracing Digital Transformation is not merely about adopting new technologies but fundamentally rethinking the organization's identity—from a traditional entity to a software-first company. This shift enables organizations to stay competitive in rapidly evolving industries by leveraging digital technologies to innovate, streamline operations, and create new revenue streams. Specifically for AI, the chapter highlights that Digital Transformation lays the groundwork for AI readiness, allowing organizations to seamlessly integrate AI technologies to further enhance operational efficiency, personalize customer experiences, and drive growth. This transformation is crucial for survival in the digital age, as it allows organizations to adapt, innovate, and thrive amid the challenges of Digital Darwinism.

Structured Data for AI

Case Study Two: Public Transit Agency

———

Imagine you find yourself in a megalopolis for the first time. Finding your way around a city can be a headache, whether you're stuck in a traffic jam, using public transit to get to work or run an errand, or you're a tourist visiting a new place. The frustration associated with each of these situations is different, but the physical infrastructure of the city—roads, waterways, modes of available transportation—plays a huge role in how much of an ordeal it is.

In order to demystify AI, it's easiest to understand how AI can derive value from *structured data*. The sophistication and complexity of real-time mobility provides us with a wealth of relatable data and wonderful opportunity to explore one of the basic forms of AI—the prediction engines on top of structured data just like the one we built for a major metropolitan transit authority.

For this case study, a metropolitan transit authority approached us to help optimize its systems and improve the experiences for its riders. As such, in this chapter, we will illustrate:

- Basic AI principles
- Architecture's important role
- Implementation and automation approaches
- Unexpected benefits of "bad" predictions

AI: No Magic

The definition of AI has shifted over time, and we don't expect this to change. In the early 1990s, the internet was still nascent. There was no Google, no Big Data, no personalized ads, no clouds, no smartphones, and no social networks. Email was an option for only a select few. There was no digital photography—and no deep fakes. Our lives were simpler and more "real," meaning less digital. The prospect of AI felt unreal—akin to magic. University students in tech major programs might have learned about theoretical clusterization algorithms, which in the best case could be used in some scientific environments like theoretically determining where to drill the next oil well. In the modern world, you would categorize this type of effort as data mining, but in pre-AI times there was not that much data available. And with no data, there is no magic.

The early 2000s—during the era of what was known as the "Business Intelligence Wars"—introduced the world to the emerging field of Big Data. Celebrating the breakthrough of 100 terabyte data warehouses would today be considered celebrating miniscule data, but it was a major accomplishment then. Data started accumulating from diverse data sources such as internet-connected cash registers, ATMs, TVs, or weather sensors. The "intelligence" then was the ability to look back in time and get an understanding of success, failure, progress, or lack thereof, from charts, tables, or dashboards. The data was almost entirely structured and had to move through various stages of cleaning, extraction, and loading into more sophisticated central systems from which the "intelligence" could then be gleaned. Data mining and Business Intelligence were often used interchangeably but were in reality not tightly interconnected. Unstructured or fast-moving data could not be used by those systems. The ability to predict the future in a nondeterministic way was still magic.

Nowadays, there is no shortage of data. There is ample compute power to process it to predict previously unknowable outcomes. We have entered the era of true AI. We may not be aware of how deeply, but AI is now all around us. We have even—in a very limited capacity—used what is called "generative AI" in the production of this book, a concept we explore in chapter 5. The lists of takeaways at the conclusion of each chapter were generated by a generative AI. We did this to exemplify AI's capacity for scanning and interpreting textual knowledge and generating content in a clearly defined format. Each illustration for the book was created by designers using multiple AI tools. The book was very much written by humans, but having AIs at our beck and call to enhance certain elements almost felt like magic.

We will not delve into highly technical, scientific explanations or definitions. Our goal is to use relatable real-world examples to explain what it takes to build an AI, what's under the hood of an AI, how we use ML, what problems we have encountered (you probably will face them too!), how we overcame them, and what outcomes we achieved. By the time you may be reading this, some of the technologies, tools, or models may not even be considered AI anymore. But today we want to keep things relatively informal and will refer to many of them simply as "AI."

There are many branches or subsets of AI. The breakthrough these AI branches have in common is that huge, fast-moving data volumes combined with virtually limitless compute power have allowed software engineers to start moving from traditional, deterministic software engineering to the more fuzzy, innovative, and extremely powerful probabilistic prediction modeling. Traditional software engineering is normally rule-based, where predefined workflows lead to predefined outcomes. Some of them may appear like AI to some users, but they are not intelligent. You have interacted with many of them in the form of some of the current chatbots or even in some of the deterministic answers you receive from voice assistants like Siri or Alexa (over time, they will of course try to increase their "IQ" by being able to better handle more nondeterministic questions). It is well known how a traditional system should behave in certain situations. If something doesn't work as expected, it is probably a glitch, and your developers should fix it as soon as possible. By contrast, any true AI is probabilistic. The outcomes are not 100 percent certain. The AI will typically tell you that most likely the result is this or that. Due to the volume of the data, the patterns it recognizes inside of the data, and its ability to extrapolate the results during phases of uncertainty provide good results in most cases—results that are good enough for us to move forward with increased confidence. In all cases, when you enjoy the results of an AI engine, never abandon your own critical thinking because there is no magic.

Predictions

To get one from point A to point AI, in most situations we encountered in organizations large and small, old and new, we faced two choices. It was either that Digital Transformation was the prerequisite for attaching cognitive services to the portfolio of the technologies the organization already had, or the organization was AI-ready, and we built AI capabilities from scratch. Either way provides a path for going from non-intelligent to intelligent.

AI eliminates or automates some of the mundane tasks in your routine so you can accomplish tasks faster. It's true that AI is grounded in automation and that there is a lot of automation happening in our world. Developing and programming is itself a form of automation, so AI is a natural evolution of programming—it's the next step, designed to give you more freedom.

Additionally, AI gives you the ability to put your data to action—sometimes in a faster, more affordable way but always more intelligently. In the past you had to have data centers and teams of people extracting patterns from your data to extract the data's value. A properly trained AI model can assist humans when they encounter problems that are not so obvious. When we don't know the best way forward, AI helps us determine the solutions because it has learned from the data on your behalf.

AI doesn't start out as intelligent, but it is a relentless learner. It depends on humans teaching it to improve. Without human help, AI doesn't know what a good or bad result is, and it cannot learn on its own from the uncategorized data. We humans *make* it intelligent by providing "good" and "bad" examples. We improve it by feeding it more and more accurate data.

Because AI has to learn, not every result will be perfect. The predictions may not always be accurate and can even at times be completely wrong. Your ChatGPT responses can sound like complete nonsense. Your smartphone photo albums' AI can sometimes mistake your child for you because you look similar. But if your solution is well architected and provides the ability to learn from mistakes, by either verifying with a teacher or against additional data, your AI will become smarter.

Predictions are one of the key aspects to understanding how many AI programs work. Predictions rely on models derived from, hopefully, accurate historical data to become intelligent. For example, you see this type of predictive AI at work whenever you use a mapping app in your car as it iteratively predicts your arrival time based on both historical data patterns and real-time conditions, such as road closures or accidents, feeding the AI.

Predictions give you a higher probability of understanding the future over time. You avoid being in reactive mode where you might miss projections and say, "What am I going to do now?" Over time, they empower you to switch from reactive mode to proactive mode across more and more business functions. Predictions are a critically important component of your organization's AI-driven Digital Transformation to prepare for the future and whether you will become extinct or incredibly successful. And the easiest approach to understand how these predictions are reached is through structured data.

Patterns

Tens of millions of people rely on public transportation for their commute and other important tasks every day. For transit users, deviations from the normal schedule—a bus arriving ten minutes late, a route change because of a construction detour, or a train breaking down while they are onboard—cause major headaches. These unexpected changes in patterns can be detected in the real-time data flows and communicated to riders to avoid confusion and frustration. Permanent data analysis is an opportunity to uncover low-hanging fruit for system optimizations that maximize the riders' satisfaction. Many cities strive to optimize the balance between personal vehicle and public transit commuters to improve the daily lives of its residents. In our hometown, Washington, DC, the local government is on a mission to reduce commuter congestion, annually passing new regulations and mandates, adjusting building codes to reduce parking construction, mandating reduced employers' parking benefits, to reduce traffic and pollution.

More accurately predicting arrival and departure times across vital pieces of public infrastructure becomes increasingly important to avoiding disruptions to businesses and personal lives. Think of how much human stress would be reduced if the system could rapidly predict and communicate public transit deviations so that riders can make other arrangements.

As a team, we always look for possible edge cases for applying innovations such as AI. If we can solve edge cases, we can solve any other cases in between. Our client, a public transit system serving one of the largest metropolitan areas in the United States, is such a wonderful edge case for predictions. Its customers are millions of riders in thousands of constantly moving buses, light rail trains, water taxis, or other modes of transportation, operated by thousands of staff members. With external forces like driving behaviors, city improvement projects, or weather, this dynamic system of systems requires multi-vector predictions, and can experience "the butterfly effect," where a slight change in one system affects every other system. Applying reliable real-time predictions under these circumstances is a satisfying challenge, with vast amounts of structured data to build countless outcomes from, scaling over time!

The complexity of such a physical infrastructure also mirrors the required sophistication of the technical infrastructure that eventually will need to be established over time to accommodate its prediction demands. For example, each public transit vehicle flowing across the system represents a real-time data stream that needs to be tracked, stored, analyzed, understood, and then used to *predict* its future states. One static schedule published every few months simply will not serve the real-time optimization needs of the real world.

Our journey with this client started with the creation of various mobile applications for the transit authority to improve the experience for the riders. Essentially, we gave the consumers the ability to find answers to their three most common questions: "How can I get to where I am going?" "Where's my ride?" "Which other transit options are available in the area?" We soon realized that the data and predictions we received from the existing backend systems needed to be improved. The information was initially not accurate enough; there were outlier cases where buses were reported being in Africa or positioned over skyscrapers! Data was not served to the apps quickly enough, even though the state of the transit system was obviously evolving across many vectors in real time. The app might show a rider that their bus was about to arrive when in reality that same passenger just saw that same bus depart. Of course, physics teaches us that no system is ever truly real time. Whatever data point was acquired this millisecond is already one millisecond in the past. Hence the importance of accelerating data streams as much as practical and feasible. Algorithms had to process the data streams quickly enough to always predict the future state of every element of the system.

Every modern application calls home by calling a specific or a combination of APIs, where the backend systems send the data. We improved on these results by inserting a real-time ML infrastructure to complement the existing backend. It consumed and processed the data streams to expose the predictions through APIs hundreds of millions of times a day. Such an improved backend leads to faster, more accurate predictions, a better managed transit system, and ultimately happier riders.

Every organization has such edge cases, where predictions provide tremendous value. In this case, it's on-time performance. Accurately predicting bus arrival times in real time is a critical aspect of making a transit system smarter. Buses must be punctual. In addition, buses literally represent the end points of the system, the edge of the network. That's where adding predictions has the most impact. Therefore, that's where the AI journey had to start, and could be built upon. The lesson here is to think big but start small and focus on what is the most actionable.

Once you successfully complete the first step of adding early-stage AI, in our observation, many additional benefits and opportunities emerge. With this new infrastructure complementing your existing systems, new insights are easier to derive at reduced cost and time investments. Your data is better categorized, the data pipelines have been laid, and the prediction tools have been deployed and tested. Looking at the frequency of bus arrivals, we detected an unusual phenomenon. We called it "bus bunching"—multiple buses concurrently arriving at the same stop with a larger break until the next bunch would arrive. Imagine waiting for your delayed transit vehicle only for it to arrive along with several other vehicles on the same route. The first bus is full, the other buses are empty. Preventing bus bunching was an unexpected and highly valuable benefit for the client. You will find that implementing AI often leads to wonderful, unexpected conclusions that incrementally benefit the system.

No matter how much effort you put into creating the perfect prediction engine, sometimes predictions are off, at times extremely so. These anomalies are your opportunity. Your data science team can analyze whether these incorrect predictions indicate pattern changes of

AI DRIVEN

your actual physical system. Some anomalies are unavoidable noise in your data, like lost or incorrect GPS coordinates in our case. Those and many others like this are easily fixable, but some are systemic, requiring analysis and deeper understanding. In some cases, these anomalies require interventions in the real world, not just the digital data world. In our client's operations center, teams monitor buses 24/7, and if some buses unusually get off schedule, the system raises the flag to investigate. Sometimes, buses are not on their expected routes, sometimes they go in the wrong direction, sometimes they are not on the right side of the road. Regardless of the issue, human involvement is required. AI assists, humans work, and now more efficiently. Incorrect predictions do a great service by systematically highlighting anomalies to the operators. Not all bad predictions are bad predictions!

Just like for our customer, the combination of pattern detection, accurate real-time predictions, additional freebies, and the identification of problems through investigating real-time anomalies will present an opportunity for any organization to improve, optimize, and sometimes eliminate some of their internal procedures and operations. Predictions can only work if the data contains behavioral patterns of the analyzed system. No one person can monitor and study several thousand buses at once. AI plays the role of the assistant. It doesn't care if it has to monitor one thousand or ten thousand buses—it processes all of them without needing a lunch break.

Architecture

Architecture is the most important concept in any IT effort. It can be broken down into three levels: application, infrastructure, and—if you are in a complex environment—enterprise architecture. You ultimately need different expertise to work on these different layers

40

of architectures. Their commonality, however, can show you how different components of your overall system of systems work together, revealing their relationships and interdependencies. In most cases, the architecture reflects your organization's structure—the collection of all functions, tasks, regulations, and future needs. Regardless of your business, architecture will evolve, and AI readiness will be part of that journey.

There are many differences between good and bad architecture; one of them is that good architecture is extendable and *future-proof.* Only you know how your business can evolve in the future. Good architecture is built to adapt to both foreseen and unexpected changes cheaply and quickly. If your architecture is rigid to adapting to upcoming changes, it is bad architecture. A simple test as to whether your architecture is good or bad is how much time and investment is needed to apply a change.

There is no such thing as perfect architecture. You have to be cautious not to over-architect and enforce economic viability. If it can't react to upcoming changes fast enough, you will be in trouble. If you invest too much into designing it for the future, the future you planned for may not happen. Regardless of how much effort you put into building good architecture, it will only be good to a certain degree. No one architecture is perfect, but all should be future-proof.

Typically, on the application layer, one or several business processes are implemented. Sometimes, it is one big monolith which does it all; sometimes, it is a microservice, which is responsible for one logical slice of your business or technology. This is the level that may or may not yet support the data exchanges with other applications or services, but it does some work, simple or complex. It's the level that contains your business logic in terms of specific functions and tasks.

At the next level, infrastructure architecture is made up of the systems that support these applications. This could include things like your security services protecting the private data stored in your various applications. With the appearance of cloud computing, more thinking has gone into the infrastructures where microservices can play a role in scaling the power of the applications.

Above this is enterprise architecture, where there needs to be an exchange of data between the various applications. Here, each application still retains its primary function. The decisions at this level look at the entire structure of the organization and its goals, influencing what kind of applications, infrastructure, or systems will drive success.

This can become incredibly complex as it contains many components, from payment services to HR systems, to customer relationship management systems or global supply chain management systems. It is important to look at each level and dependencies to compare your status quo and desired to-be state, so you can strike an economic balance between doing too little or too much.

When approaching these three architecture levels with AI-first thinking, you earmark system components or workflows that can be improved, then automated with AI in the future. It's OK to put temporary placeholders in the system knowing that once you update other parts of the architecture, you can build intelligence on top of the existing infrastructure. These placeholders should follow a certain cadence. In our customer example, at first, humans observed the incoming location data in read-only mode. As the data became clearer, we allowed manual intervention to the schedule. For the next semi-automated step, AI assisted the humans by suggesting schedule updates. Finally, it relieved the humans by only alerting them when it detected system anomalies in real time.

For this effort, we needed to integrate our predictions and the various legacy customer systems. To accomplish this, we only had to add one piece to our transit system, not rebuild the whole system. On the customer's side, the one part of their system that was future-proof was the data ingestion module. Unfortunately, the existing prediction infrastructure was poorly architected such that the alternative to a lightweight integration of the two modern modules would have been a complete re-architecture and reengineering of legacy systems. We inserted a prediction module into our transit technology system. We were consuming the data anyway, so it was easy to spin off another data stream with an inexpensive and easy implementation. The prediction results were exposed back to the customer. On the client side, they simply added one more data feed intake to their systems. As a consequence, and at a fraction of the budget that would otherwise have been needed, the experience for transit riders was dramatically improved through accurate and timely predictions. Without such modern architecture, we wouldn't have been able to integrate so easily.

We started with one metropolitan area, processing the data generated by several thousand vehicles, but architected in a modern way that allowed us to onboard others without a delay, whenever we wanted to. At the time of writing this book, our system has evolved to analyze more than seventy transit systems, making hundreds of millions of predictions daily without any meaningful increase to the annual budget.

This is why architecture is so important. It is where true efficiency can be achieved and where you control the economic variables of your downstream integration and engineering efforts. Ignore the architecture and disrespect the data at your own risk. Architecture determines your AI readiness.

Black Box

At the beginning of an experience like with this customer, it's typical to encounter older monolithic systems that are vendor-proprietary. We call these systems a "black box." Closed and opaque systems like this make analyzing, understanding, and sharing the data and prediction calculations nearly impossible. They are also error prone. Oftentimes the only approach to work with these systems is reverse engineering, which is painful and can be prohibitively expensive.

Our approach was to focus on enhancing the infrastructure through nimble technologies—cloud services, cognitive services, APIs—to make the right data available to the data consumers. We were able in this case to open "the black box," so everything changed from being a mystery to becoming openly accessible for scrutiny by anyone to check it, use it, learn from it, add to it, and design new innovative solutions.

An additional benefit of opening the "black box" means different pieces of the system can now be placed nimbly in the most appropriate infrastructures. One of the goals should be to minimize recreating the software services you can "rent" as needed from reliable, openly architected expert systems. Creating intelligent scalable cognitive services is not trivial. It is also hard to achieve multiregion interconnected data center networks with all their hundreds and hundreds of offerings, only one of which is the AI products we just mentioned. Shared resources like this across millions of organizations are generally cheaper and better than anything you might be able to create on your own. As for cognitive services, there are plenty of enterprise-class options available on the market such as AWS, Google, Microsoft, or OpenAI, with its recently released ChatGPT. By the time you read this book, there may be many more—perhaps even better—ones available. Try to avoid getting locked into a "black box," or you will never be able to take advantage of all these exciting innovations others are doing on your behalf.

To prevent accidentally stumbling into the next "black box" scenario, it is important to overlay these kinds of systems—open or closed, legacy or modern, intelligent or dumb—with a set of open APIs, and make these systems exchange their respective data through APIs and the data exchange contracts they enforce. This allows you to add, remove, and modernize systems at your pace, while opening "black boxes" when needed. Unlock your data, and you unlock your innovation capabilities for others. The power of the predictive engine we created for this customer ultimately emerged in countless applications that utilized our AI through our APIs. Suddenly, every transit vehicle across this vast metropolitan area became visible in real time with accurate predictions of their arrivals, not just our mobile apps— where our journey with this customer started—but also in Google Maps, third-party transit apps, and local publishers' websites.

"Black boxes" are dangerous black holes in an organization. They zap your budget, your creativity, your openness, your velocity, your ability to integrate internally and with others, your chance to generate alternative revenue streams, and therefore possibly your ability to survive in the innovative AI economy. There are proven methods to shine the light into them and to turn them into open systems like the ones we discussed here.

Optimize

The one thing we *did* know about this client's problem was that we needed to optimize its ability to generate and manage more, and more accurate, data, in real time. To make the predictions more precise, we needed to look at the data infrastructure itself, how the data flowed through the system, where and how it got stored and processed, and how the results were exposed to other systems and consumers. We needed to build more models on top of the existing data, while designing new data flows to discover insights the previous black box was incapable of providing.

As is often the case, most of our technical optimizations in this instance were driven by economic considerations. For example, there are different types of data storage—cold storage, hot storage, in-memory storage—all of which have their unique benefits. Many companies still have a central data warehouse for storing, analyzing, and reporting their data, regardless of whether their business needs have diversified away from purely reporting historical insights. Others still keep their data in a database; however, if you're dealing with real-time data like this client was, then such a scenario will fail in a very short period of time because you'll run out of budget or performance—most likely both.

We needed more advanced data architecture to optimize the budget inevitably needed for the ongoing support of the resulting infrastructure. The prediction results had to be generated as fast as possible for which we required the fastest storage possible. We also wanted to be scalable across other jurisdictions because we don't like repeating ourselves. The technical decisions had to be optimized across three axes of a classic decision triangle: Budget, Performance, and Complexity. It's hard to satisfy all three. The decisions had to be balanced, taking trade-offs and interdependencies into consideration—a solution to the customer's problem and needs that was as economical as possible while accommodating the highest feasible sophistication at the fastest necessary speed.

As it accumulates relentlessly over time, historical data takes up the most space. In our case we needed as much of it as possible to train the prediction models, and to maintain an audit trail for regulatory purposes, which requires keeping at least half a decade's worth of data. Yet, besides training the models and other purposes, we didn't use this data often. The volume of data was enormous, so we used the slowest, least expensive option provided by the cloud provider. Each trip within the vast transit system was stored in cold storage during the day, and used to retrain and refine the model overnight, when fewer buses were en route. This economical option was sufficient to feed the necessary sophistication of our prediction models while never becoming a bottleneck for the system.

Most modern ML frameworks work with the data in-memory, and so all real-time data required to make our predictions was additionally put into memory. The in-memory data was used exclusively for making predictions and storing their results. There were no additional technical layers between end users and consumers, allowing us to deliver the information as fast as it could possibly be done technically. Speed of intelligence delivery was paramount here.

On a gradient somewhere between slow, cheap cold storage and blazing-hot, sophisticated, "expensive" in-memory storage, we also inserted a traditional relational database, which contained regularly updated, relatively frequently used data for agency operators. Slowly changing data, such as all the drivers' and fleet information, was stored there, like the location of bus stops, and other data that is required to intelligently manage an agency like this.

The various technical optimizations lay the foundation for enabling evolutionary—sometimes revolutionary—business optimizations. The ability to get an accurate sense of when and where more or fewer vehicles are needed at any given moment, and how to optimally adjust the vehicle frequency at a moment's notice should be expected from such a transit AI. The transit agency could now create a radically more accurate schedule to set expectations with riders as to when to anticipate their buses, at what stop, and on which route, at any moment in time.

Now that the schedule mirrored the realities in the real world more precisely, it taught the models, enabled further improvements, and created a virtuous loop of constant improvement. Our AI had gained the ability to correct unforced human errors! We could predict what route a certain vehicle was on, even if a human made a mistake and entered the wrong data. A driver may have accidentally pressed "route number 32 Northbound" instead of "route number 23 Southbound" to display on the in-vehicle and transit stop systems, potentially confusing riders, the operations center, and others. Yet the anomaly detection abilities of our AI picked up the discrepancy between the driver's error and the factual data in real time, and allowed the system to accurately reflect reality before any mistakes could even set in. This example of the power of technical optimizations leading to possible innovations in the real world is considered revolutionary in the transit space.

Iterate

You may be a parent or at least have experienced babies at some point in your life. Remember when they were first born? As a parent you have a range of different feelings: happiness and excitement, as well as worries and sleep deprivation. Your baby is the most important human in the world to you at this time. Their first interaction with this world starts with the repetition of very basic tasks—eating, sleeping, pooping. Most of these are basic instincts, they are not learned. The reactions to external stimuli are very limited. Initially, the infant only reacts to contrast, bright or dark, without true focus on anything for the first few weeks. Then you notice that their eyes stay longer on some objects (image recognition!). Their vision is evolving. Eventually, their eyes follow your cat or dog (object tracking!). Then they start recognizing family members (face recognition!) and give you a huge smile every time they see you—probably, one of the happiest moments for you at that time.

Like infants, AIs start out as a blank slate, and it was no different in this client case study. Every future prediction was poor in the beginning. The key was to advance and optimize after the expected initial failures. Decision-makers had to appreciate that the ML engine was learning and improving its predictions over time. Today, we can predict the entire transit system's many moving parts past thirty minutes with extreme reliability, which is an eternity in the scope of a constantly moving transit system! No one wants to stand at a bus stop waiting for half an hour, so we had scaled the system to the point where it could predict everything it needed to predict within that time frame.

We started small, attempting reliable predictions three minutes out. Once the data showed we achieved more accurate results than previously used methods, we added and tuned incremental prediction intervals of six minutes, ten minutes, and so on as the predictions improved.

Incremental iterations give you the freedom and opportunity to experiment, learn, and apply the learnings in a continuous process of improvements, such as performance tuning, at low risk. It's about becoming better today than you were yesterday.

Creating a disciplined iteration strategy is key for this kind of success. If your expectation is to have perfection right away in one "big bang," your likelihood of success is zero. Instead, you design a system in which you start as small as you can and then establish regular iterations. Assuming a cadence of weekly iterations, each week you prioritize and act on something such as moving a product feature, a system enhancement, or a prediction result into your production system, where your customers interact with you digitally. Then your job is to focus on making sure there is a positive rate of change for each iteration that stays constant week after week.

Let's say we improve our predictions with each iteration by only 1 percent. The continuous improvements compound upon each other, creating a hockey stick effect. Before you know it, you reach prediction Nirvana much faster than if you tried the "big bang" approach. This disciplined method reduces your risk. If you have a week of iteration without improvement—or the predictions are worse—then you've only wasted a week's worth of time and budget within a multimonth, multiyear, or preferably infinite process. But you've learned from it and can make up for it in the next week's iteration.

This process of iteration allows you to accumulate the right data—the raw commodity from which your AI will iteratively be able to extract your unique currency, the specific insights into your operations that might provide an incredible business opportunity to you. This iterative process enables you to continuously experiment in a rapid cadence. It allows you to maximize your organization's return from these insights, be it more satisfied transit riders due to more accurate predictions, reduced CO_2 emissions due to a more efficient transit system, more revenue, less operating cost, or whatever other KPIs matter to you.

Therefore, it is imperative that you learn how to keep your data, control your data, and share your data on your own terms. A short time after we set out to write about our experiences here, one of the many controversies surrounding ChatGPT involved content publishers, whose proprietary work had become just another data source to train a model for just another AI. Use the learnings you gain from this iterative process on how not to become just another data source for someone else's AI. It either must be your own AI, or someone has to pay for your data.

Practically, what this means is that you don't want to move your data closer to the applications, but rather iteratively move the applications ever closer to your data. With this client, we began with five years' worth of historical trip data. That's a huge amount of intricate data to move around. In the past, many systems were systems of record. To create new systems to integrate with, the data had to be extracted and moved from other systems to the new systems. The amount of data was small and slow, so the approach may have been valid. Think of ChatGPT again. The entire internet is its data source. Moving petabytes from one system to another is not feasible. For one, the internet is simply neither fast nor cheap enough. Instead, you allow others to plug into your data, and architect your apps to plug into others' data to be able to play with it at its source. College students do their research in the library, where the books already are, rather than lugging all the books they need all the way back to the dorms. Your apps should do the same over time.

There's an old saying that no business plan ever survives its first day in the real world intact. Likewise, you have to be flexible in the assumptions you're making as you are entering your AI journey, get in front of your potential prediction consumers as quickly as you can, and iterate from there in disciplined cadences. Expose your assumptions to the real world and learn from your mistakes. Remember when your infant went beyond its basic instincts to learning new skills like sitting up. First you assist them, of course. Yep, they can keep their balance for a few seconds, but then, boom, the left side is heavier than the right side. We try again, and yep, another few seconds, nope, the right side is heavier this time. We try, we fall, we try again, we fall again. We try the same routines for a week or another week, and at some point, the skill is acquired reliably. After thousands of attempts, your child can finally sit on its own. Even if you nudge them gently, they keep their balance. We sit!

DevOps

We have all experienced it. You open your favorite mobile or web app to get information, play a game, or make a change, yet instead you receive a message "the app is down for scheduled maintenance; please try again later." Annoying, isn't it? Less annoying, but still frustrating, is to open the app, only to see outdated or inaccurate information or wait and wait while the app loads the data. While architecture plays the main role in performance, DevOps is often the main driver for reliability and accessibility.

There is no strict definition of what DevOps is and what DevOps is not. In our world, DevOps is a discipline that is responsible for managing infrastructure by automating, monitoring, deploying software components and infrastructure updates continuously. A true DevOps engineering team's goal is to make managing the infrastructure less dependent on manual interventions by automating everything.

There are new trends, tasks, optimizations, systems, even regulations that have to become part of your solution as well. Constant changes are inevitable and have to be accommodated in DevOps practices and policies. The faster you deliver your changes to your production environment, the faster you validate your changes with your customers. Their feedback allows you to identify and fix any issues or adjust your vision. This ability turns into your superpower against your competitors, whose reaction time is slower than yours. Without DevOps properly implemented, for any—even the smallest—changes, you need your production operations team to manually build, verify, deploy, and test your entire system. Consequently, you have to notify your customers that your organization will be unavailable due to planned—or worse—*unplanned* maintenance. By today's always-on, zero-downtime standards, such manual deployment and maintenance practices would understandably be considered insane.

In its early stages the deployment of our transit technology system was far from fully automated. We delivered the solution in iterations as quickly as we could, but we had to sacrifice the complexity of many deployment automations in favor of helping the client's customers as quickly as possible. Of the three dimensions of the decision triangle—speed, complexity, and cost—speed to deployment was paramount. Everything was deployed from the development environment directly to the production system—admittedly, not a best practice. Yet, with this approach, we could react to errors quickly, because developers were extremely close to the production system. We might still accidentally miss something and deliver the developers' "leftovers" into production. After a few hiccups, we decided to focus on minimizing the possibility of such mistakes by implementing our DevOps best practices.

DevOps reduces manual work and the human factor, and as a result improves the systems' reliability. For example, implementing automated end-to-end tests is typically not a cheap effort, but it pays off after a few updates of the system. After adding the end-to-end tests to our CI/CD pipeline, we also *iteratively* added other support systems, such as monitoring solutions to understand the health state of the overall system by quickly looking at all charts. We also integrated alert systems—a proactive mechanism to detect if something is already down or about to go down—and scalability capabilities to scale down when the load is not high or to scale up the right resources to process the data we collected during the day to retrain our prediction model. All the above gave us peace of mind. With each improvement, we freed up the talents of the production support team to do more creative and time-sensitive tasks.

During the next round of optimizations, we containerized all the microservices and orchestrated them using Kubernetes, a modern abstraction layer on top of highly complex and diverse infrastructures that was open sourced by Google. Using containers meant that we could use whatever programming language we liked, because for different development tasks one programming language is more suitable than another. Without creating a "zoo of tools," this approach allowed us to break out of the "developer prison" where you force developers to adopt a certain development platform whether it is most suitable or not. Using Kubernetes gave us the flexibility to run the entire suite of environments wherever we wanted. As an experiment, we redeployed our solution to another cloud provider and verified that everything still worked as expected. As a result of this DevOps journey, we ended up with a flexible, reliable *future-proof* solution, ready for any changes, that provided a zero-downtime experience to happy riders.

Tip of the Iceberg

When you tell a video streaming service whether you enjoyed a certain movie, you probably don't see it as teaching AI behind the scenes to compile an endless playlist of content tailored for your personal tastes. Your photo app can show you hundreds of photos where it correctly identified your spouse, and then asks you to please tell it whether a handful of others also contain the same person. You probably did not perceive this as AI in action, learning how to better assist you in the future. How well its predictions work can appear like magic. When you prepare to take public transit and glance at your ride's predicted arrival time in your app, the display, or the traditional schedule, you probably didn't think much of it. Yet in each case, an intricately crafted, iteratively released, and continuously optimized software architecture is at play behind the scenes. It relentlessly consumes massive amounts of data, processes it, and continues to learn over time. Its predictions create insights and detect behavioral anomalies for you based on your needs in this particular moment in time. As a rider, you look at one number only when your bus will arrive, but it is only the tip of an iceberg of mountains of numbers being crunched by sophisticated systems.

None of the modern applications of AI were possible in the more recent past. Compute power, algorithms, and data are AI's three key ingredients. The cloud gave us the needed compute power. The internet gave us reams of public data, such as text, images, videos, even code, and other data types. Wearables, mobile devices, and many other sensors on the edge provided highly personalized, precise data. Data scientists combined compute and data to create sophisticated prediction algorithms, many of which are publicly available for all of us to apply.

To create great AI, data is the most time-consuming of the three ingredients. There may not be enough of it to train the models. It may not be fast enough. It may lack accuracy or breadth across a variety of domains. No data: no AI. Not enough data: starved AI. Slow data: outdated AI. Bad data: bad AI. With ever-increasing volumes of data, you are forced to solve scalability tasks using best-in-class DevOps for automations and data science for optimizations.

All innovations happen in production systems, the rest is research. AI cannot make any sense without end users consuming the outcomes it produces. The faster you deliver your insights to them, the faster you get their feedback, and can get your predictions more and more accurate. True DevOps is again your best friend here to ensure constant production deployments and zero downtime.

A typical rider spends mere seconds per ride using this transit AI at all hours of the day. There was no window for system maintenance. Seconds are enough time to look at a perfectly balanced schedule, allowing the riders to leave home on time, wait for the least amount of time at the station before the transport arrives as the system predicted. These short interactions of the real world with the probabilistic digital technologies rest on the complexities and vastness of many deterministic software solutions underneath it, guaranteed by a DevOps team. Like the tip of an iceberg, these elegant architectures are typically hidden, and only the AI on top is exposed to the real world.

Executive Summary

I'm a CIO of a S&P 500 company and from a different industry. I need a very short explanation (three to four sentences) of what AI is and what AI is not from the chapter above.

The chapter defines AI as a probabilistic model that leverages vast volumes of data and computational power to make predictions and automate tasks, moving beyond the deterministic rule-based software engineering of the past. AI is not a static or magical solution; it requires continuous learning and improvement, heavily dependent on quality data and human guidance to discern between good and bad outcomes. It emphasizes that AI is about enhancing decision-making and operational efficiency through predictions and insights, rather than replacing human intuition and critical thinking. This nuanced understanding highlights AI as a tool for augmentation rather than a replacement for human expertise.

AI-First Mindset and DataOps Methodology

Case Study Three: Multinational Financial Company

Imagine experiencing Steve Jobs' reality distortion field in the auditorium when he introduced the first iPhone on January 9, 2007. It was so revolutionary that it disrupted multiple industries. The iPod had already changed how people listened to music and turned the entire music industry upside down. By combining the iPod with a phone and connecting it to the internet, the iPhone became a category killer. The other "smart" phones treated the internet like a "baby internet," accessed by inferior devices and software. The iPhone as the leapfrog product presented the perfect interplay of software and hardware.

We are now presented with a similarly revolutionary concept through the convergence of broadly available cognitive services on vast global networks of computers, running very powerful GPUs, coupled with already trained or at least trainable AI models on vast amounts of data. This is already starting to change the way people complete certain tasks just like the iPod changed the way people listen to music and thereafter disrupted established industries. Just like Blackberry devices were defeated by the iPhone, as of this moment, in an internet-second, the industry disruptor Google is now defending itself against ChatGPT, "the new Google." Any company and industry that has not put itself into a position to embed tools like ChatGPT into their systems will follow suit and be disrupted by those that are AI-ready. It is increasingly imperative to embrace an AI-first approach.

New concepts such as the iPhone and now AI force developer communities to adapt and create new methodologies and frameworks. Over the years, we found several to be highly useful. We will use the following case study to introduce the reader to the best practices of how to start and investigate them in more detail within the context of a large enterprise adopting AI. Any organization confronted with embedding these revolutionary AI capabilities into their existing systems can mimic a similar mobile-first approach—with several unique twists—to begin and accelerate their AI adoption with rapid execution across their teams.

Applying these AI accelerators that we will explore can significantly shorten the time from inception to production, ideation to productization, and thereby the overall speed-to-market. The longer it takes for any innovation to reach the market, the longer these efforts remain cost centers without realizing their ability to turn into profit centers that can add meaningfully to the top and bottom lines of the organization.

In this case study, we focus on a prominent international financial company and how a new DataOps methodology motivated by an

AI-first mindset created an AI-ready entity. This allows us to show you how:

- To look at business needs through an AI lens
- To introduce DataOps that can future-proof an enterprise
- To identify quick wins such as Minimal Viable Predictions and Minimal Viable Processes
- To conduct an AI readiness assessment
- To execute across the organization

To Call or Not to Call

Imagine for a second that you work in a call center of a multinational debt collection agency, responsible for contacting customers to pay their bills. It's not an easy or particularly pleasant job to begin with, but it becomes even more difficult when you add several unknown factors, especially the unpredictable nature of humans.

Many businesses depending on outreach can waste time unsuccessfully repeating calls to the same customer because they are unaware the best time to call each person. Some customers only pick up the phone in the evening, or every other Monday. Some pick up the phone on the very first call, while others may have to see the same number come through three times—or more—before they finally answer. Yet others may purposely avoid calls—they see the debt collection agency's number on the screen and immediately decline the call or let it go to voicemail.

An additional aspect of the problem is that even when someone does answer the call, you cannot assess the probability of their willingness or ability to react to the service offer or the need required. Some may respond in that first call and say, "Yes, of course, I'll pay my phone bill right away." Others may require multiple interactions before they finally say, "OK, I understand. I'll pay within two weeks." Still others may never pay using this uninformed approach to connect.

These known unknowns allow for a more intelligent strategy, because if you simply call every customer as often as you can, you risk angering them or creating other negative consequences with the larger financial risk that concurrently the proportion of the unpaid versus paid bills does not shrink. Making one hundred call attempts within a certain hour instead of fifty does not necessarily double your success of getting to speak with an actual person or double the amount of positive outcomes. This approach has a high probability of failing, because you are still not aware of the right time to call each number, what the probability of a given individual is to pick up, and what their likelihood is of agreeing to an outcome.

What if you could reduce the number of calls it took to receive a payment by intelligently avoiding most of the fruitless attempts? How much more efficient *and* effective could you be? How much more profitable could the company be?

This is the exact problem faced by the company in our case study when looking to both simplify and optimize their processes for debt collection. Rather than squeezing payments out of people, their strategy is to nurture their clients back to financial health over an average of ten years. This is a noble objective, but it's difficult to achieve unless you actually connect with the debtors and accurately assess their propensity to pay.

When the company partnered with us, they had steadily grown over the years through international acquisition. Each new acquisition had its own systems, data streams, algorithms, and so on—meaning not much was synchronized or optimized across the organization technologically. Our job was to glean additional, actionable intelligence from existing data and, in recognizing the propensity of continued evolution as a steady acquiror, create highly automated repeatable processes to constantly provide insights that increased call center success rates. Ultimately, we had to ensure that they got this right—to call or not to call.

DataOps

DataOps is a methodology of best practices that helps one identify the way the data should be handled and streamlined to get the greatest value from it. It is made up of tools, frameworks, policies, practices, and methodologies all together. It includes what one needs to do to store data, extract data, process data, visualize data—and forms the pipelines between all these components of a system and how it works. DataOps maximizes the value of data. Superfluous data is ignored, until it is needed and subjected to the same discipline.

In our experience, DataOps is all about being disciplined enough to respect data. Without that respect, data will be messy and become an obstacle to wide adoption of modern innovation and fast execution.

Failure rates when interpreting data can be quite high. The CIO of a major CPG company recently disclosed to us that his organization gathers four million unique consumer data points daily and didn't know how to best deal with this avalanche of data. The CIO of another major retail company was swimming in a data lake of 50 petabytes, and he was struggling with extracting value from it. The CIO of one of the largest global pharmaceutical companies told us that she typically starts one new major data project every quarter, yet three out of the last four had been failures, and she couldn't identify the root cause.

If you don't have DataOps thinking in place at your organization, then it is data chaos. You have data, yes, but either no one knows where it is or no one knows the best way to extract, process, and get any value out of it. Also, lacking unified DataOps practices means contradicting teams are using the data differently. Your revenue team can extract the number of active users based on the number of recurring payments, because that is what they care about the most. For your engineering team, the number of active users means the number of last log-ins to the system within the last month. Both of them are right. Unfortunately, both of them are also wrong. While both groups are accurately referring to a data set, they result in different conclusions, ignoring the need for a mutual understanding between teams on how to approach the data and resolve toward an actionable result. As CEO of such an organization, you're flying blind until you resolve those data conflicts.

One of the characteristics of DataOps is having the ability to extract the insight within a relevant time frame. You may have a brilliant and well-structured data architecture, but if you get business insights a year, a month, or in some cases a week later than you need them, they have no actionable value.

The company in this chapter's case study was actually well organized around their data; it had a well-defined data approach, a good cloud-based infrastructure, a team to collect the data, and a team to process the data. It had built a data warehouse and had a data science team to find insights in the data warehouse. In this instance, control and order were well established everywhere: no data left behind and no data swamps to clean. Every piece of information was properly supervised and treated on its journey from the data source to its final consumer. The policies were declared in enforceable and well-managed data governance documents. It was clear who could see what and why, especially when data was as sensitive as personally identifiable, private financial information. There was simply no other way but to guard each data element with such protective mechanisms across all the data pipelines.

However, to maintain order all changes in the system had to be discussed, synchronized, and agreed upon between the different parties, who were responsible for different aspects of the system, then documented and finally approved. It unavoidably created internal dependencies of one team on another. As a result, each team became a bottleneck in the data flows. Yet the modern world, which is more real time than ever, demanded answers with zero wait time.

When we partnered with our client, our main task was to increase overall efficiency. As for most of our projects, we timeboxed ourselves—if you can't extract the prediction fast enough, you allowed your focus to stray and invited scope creep into your task. It became clear that using the company's existing DataOps approach, we wouldn't be able to deliver results on time. Instead, we worked with its raw data directly to explore an alternative way of getting insights quickly. If successful, we could use this new approach as a reference to modernize its current DataOps practices.

Not everyone has the luxury to change DataOps approaches, but this customer had full control and ownership over its data. Technically, it could experiment with multiple approaches in parallel, compare them, and choose the one that served its business purpose the best. Data warehousing was good and efficient here for standard reporting, but too slow for real-time predictions when someone had to make a call this very second.

While all existing DataOps practices were efficient and served their purpose well, they did not apply universally. It was not our goal to change the client's entire DataOps either; it regardless would not have been feasible within a reasonable amount of time anyway. But what was reasonable was to build an alternative approach to accelerate predictions and make sure that consumers were contacted to validate the results and document the entire process for others to emulate. The key components for successful DataOps were already in place—ownership, control of data flows, identified end-user consumption of every piece of data. The speed of reliably surfacing accurate insights was missing.

First AI—AI-First

When smartphones were first introduced, businesses attempted to retrofit their existing systems to take advantage of this new form factor. The systems continued to be optimized for desktop web browsers first, mobile second. At the same time, organizations could not ignore the fast adoption of smartphones by consumers. They attempted to adapt their desktop systems to also serve end users on these new devices. The initial idea of scaling down an existing website into one-twentieth of the screen size almost always failed. It was much harder than it originally looked to squeeze the rich content and complex layout into a limited space.

Adding new features to existing web apps to take advantage of the devices' unique capabilities, such as orientation, GPS, camera, and others, was almost impossible without big investments of time and money. Some organizations (even Apple!) failed to anticipate the need to be able to display their content delightfully on a mobile device. The revolutionary approach that overcame these challenges was to switch the order and design the systems for mobile first. Scaling up to a bigger screen is always easy; turning off mobile-specific capabilities is also "budget-friendly." Nowadays, systems are often designed for mobile devices only, and the desktop gets increasingly ignored.

Like the mobile-first approach, AI-first means that instead of adding AI elements to existing systems, you *start* by building your systems around AI. You identify a key problem or task that AI will tackle, design the system around the AI, and then build everything else on top of the core. This puts a solid foundation in place for incrementally expanding the AI footprint of your business as you add more tasks and processes for the machine to tackle.

The main idea of AI-first is to design systems where some components can use AI, even if such an AI service doesn't yet exist. Interfaces might have multiple types of data visualizations to allow end users to make informed decisions, and tools to act upon the findings. In the future, interfaces may be enhanced with additional AI prediction visualizations to make the end users' decision-making process simpler and smarter. It could go even further and automate safe decision-making processes. At some point, as additional AI capabilities prove valuable, the dashboard may not be needed at all and will be replaced by a fully automated AI process in the backend.

This possible future state of your system to run your business will be reflected in a "to-be" architecture, which your engineers will have to work toward from your current "as-is" state. Architects therefore need to appreciate that some components will act as placeholders in the beginning, and although they may never be replaced with AI, the opportunity is not ignored in the event the business requires it. Good data scientists are constantly asking themselves what kind of data they may need to gather, process, and make predictions on—or not. Such data is potentially unavailable or may never be. But they should be able to think far enough ahead about what the future can look like. For designers it means that if they design the interface with "AI in mind," the human may continue doing this work manually, or semi-automated until the AI is ready to replace some of their manual steps with new, more automated, and enhanced UI tools. Ideally, over time, all functions, not only within your IT department, are trained to continuously think AI-first even if it is not yet or may never be inserted into their processes and tasks.

Our financial client collected all the data it possibly could across the various countries where it operated. It knew how to process the data to build useful reports and now was trying to add predictions on top of those existing systems. This resembled the way desktop-first applications were squeezed to fit into mobile devices instead of the applications being architected and designed around mobile first. The systems and processes at our client were not ready to adapt to new AI capabilities. It proved as hard to accomplish as retrofitting a desktop application to be mobile-friendly. It can be done but takes time and money with no guarantee of sustained success.

The customer attempted to create one unified view of everything all at once on top of all the data gathered, which effectively acted like a systemic data hoarder trying to analyze *all* of the data to derive actionable insights from it. Our plan was to slice away some of the clean data and build a statistical model on top of it. Our approach was to choose one country, get a deep understanding of the untreated data, and build an accurate prediction engine for it. Call centers are unique in each country as they each have different regulations, cultural norms, and data that can or cannot be collected. Instead of assembling and then disassembling the data, we worked directly with the raw data, specifically the particular slice needed to make the first predictions. Should one decide to add more, different predictions in the future, it would be easy to replicate the whole data pipeline after it was first established. In our case, this worked and became the approach we used for subsequent countries. It did turn out that individual country citizens differed from each other and expected to be served differently by their financial service providers. Patterns vary and regulations are different, yet the approach to building an AI system remains the same.

As a best practice, we kept the data at the customer's original source, stopped moving it across borders, and instead moved the apps, models, and predictions closer to the data origination source, exposing only the necessary results with the best available tools. We ignored all superfluous data, adjacent systems, and stayed laser-focused on only the data we needed, extracting the highest value results with the simplest possible prediction AI on top of it. The rest of the systems and integrations of the customer now had a reference architecture and proven path to AI it could replicate, enhance, and adapt to as needed. The AI seedling had been planted, allowing it to spread throughout the rest of the organization in the most organic way: AI-first leading to the first successful AI implementation, so others could follow and do so more easily.

MVP, MVP, MVP

When it comes to developing a software product, the concept of a Minimum Viable Product (MVP) has been well established. It contains the most minimally viable set of features and functionality, providing the most concentrated usefulness to the product users you can later build upon in subsequent iterations. *Minimum* in this context means the least expensive and fastest way to get your product to market. This allows you to gather feedback, clarify assumptions, and create a technical feature, functionality, and business road map to give you the highest probability of continued success with your existing user base. Along the way you will gain the understanding of how to reliably attract new ones. *Viable* allows you to validate your idea in the quickest and least expensive way regardless of whether you cancel the development efforts before they become more time and budget consuming. The *Product* identifies the smallest set of features and user functionality that you can build within the smallest investment of time and money while still getting real user engagement.

Two other notions of different MVPs—Minimum Viable Prediction and Minimum Viable Process—that we have created over the years are important parts of our know-how that allow our customers to validate initial hypothesis of implementing AI. One identifies the prediction with the highest value, while the other creates a shortcut from data science to production. The three MVPs overlap and strengthen each other's utility. Each is distinct from the other and serves a separate purpose to accelerate the execution of AI efforts.

When it comes to whittling down a set of prediction models, the same principle as MVP—understand their potential usefulness quickly—should be applied. In the case of a Minimum Viable *Prediction*, the first users of a Minimum Viable *Product* are the first beneficiaries of the most viable prediction results. Unlike a software product, predictions have an unemotional nature to them. Their nature is probabilistic; they are either accurate or not. They either have a higher probability of turning out to be true than not—or they simply have not proven valuable at all, so you can move on to the next one quickly.

Unfortunately, prediction models also often have a Black Swan nature to them. They work correctly until they don't because environments, behaviors, usage patterns, or preferences change constantly. Fixing the models can take a big effort of assembling different data, adjusting data processing, or increasing data quality or velocity. Otherwise, they have lived out their usefulness completely and need to be replaced with something else. Even when predicting the future—should I call now or later?—unforeseen and unforeseeable changes in circumstances are an inevitable part of the effort to prepare for. Due to these complexities and real risks of getting a prediction AI wrong, it is imperative to keep your first attempts as minimal as possible, so you are not yet required to overinvest in data pipelines, compute power, infrastructure, or ML and data science skills. To accomplish the minimalistic aspect of this MVP principle, you must identify the simplest business question to answer. This increases the probability of models to predict an accurate answer with the data you have. This is your quickest path to finding the truth.

In the case of our client, they had bought billions of dollars of portfolios of unpaid debt from a variety of markets including delinquent mortgages, missed car payments, and unpaid phone bills from telecom companies at an attractive discount. The constant question remained how to collect more from each delinquency above the purchase price to meet shareholder return on capital requirements. Ideally, we wanted the AI to predict the best communication strategy for each individual debtor to recover the highest amount of debt overall. For each of the millions of individual debtors, the AI would recommend the right communication frequency, the right communication channels, and the right tone for each message. To increase the level of personalization, it would take into account the correct amounts to request, the various country and regional mandates and laws, and cultural norms. None of these variables are simple business challenges to address with prediction models. Your organization may spend much engineering capacity and invest countless dollars yet come up empty-handed unless you pick such a minimally viable path first, and then iterate based on the first successful results.

A Minimum Viable *Prediction* requires a disciplined approach to patiently remove every layer of complexity from questions such as these. Every business like the one in our case study naturally wrestles with this level of sophistication. Instead of considering all possible ways of customer communication channels, we chose one that already existed, such as this client's call center, so we didn't have to wait until a new channel had been architected, engineered, integrated, tested, and deployed before attempting to predict its viability. None of the individual debtors was like any of the others. Some will pay, some will not, some will pay a fraction of their debt, some in increments, some will pay regularly, some irregularly, the variations are as numerous as the number of the individuals who owe the debt. The purpose of this entire AI project was to maximize the total payments and increase each individual's payment.

The complexity of such a highly specific customer communication strategy requires an equally intricate model, trained on enormous amounts of data. So instead of taking into consideration all the aspects we ideally wanted to predict, we minimized the complexity by removing one variable after another. As a result, we arrived at the most minimal, yet still viable, task that our Minimum Viable AI engine needed to predict. When your business is about connecting with your customers at the right frequency to increase the number of successful transactions—such as in this case—the binary Minimum and Viable prediction to make is Yes or No: Should I call this particular customer now?

We enabled the call center agents to not only reduce indiscriminate calls but to also increase the probability of success for each attempt. They should be the first group impacted by the results of the MVP. There are many predictions you can probably envision for call centers. But the most impactful one was for the AI to advise the individual agents when to call and when not to call. Minimum Viable Predictions automate and predict the most repetitive task or situation you identify that has the potentially highest impact. To call or not to call was identified as the customer's Minimum Viable Prediction, and the validity of the model was easy to assess in terms of successful call attempts and payment growth rates versus more robust pre-AI.

There is never a guarantee that the model's prediction of that first variable—to call or not to call in this case—will have a positive impact, a negative impact, or no impact at all. Intuitively and based on our training results, it usually should. That's why we selected it as our most minimum, *viable* starting point. We were in a position to measure the results rigorously and make one change at a time while keeping all else constant, such as the number of phone operators, the hours they worked, and the message they delivered. We entered the typical period of constant tweaking, experimentation, and thoughtfully adding complexities in the form of other variables that could positively, negatively, or fail to compound the initial results. The next variables that could be tested were a variation of predetermined call scripts and other additional layers of complexity. The AI would be able to predict not only the right moment to call but also what type of message to use to improve the probability of the debtor paying their outstanding bill in full. During this evolution, more and more levels of complexity were gradually added to refine and extend the model with more targeted, precise, compute-heavy predictions that measured the additional successes again and again.

No prediction is ever valuable without a real-life outcome. For that, your engineering teams need to identify the shortest path from the system that generates the models' results to a production environment that the consumers of the intelligence—in our case the call center agents—can access. Your desire might be to tightly integrate the prediction into existing systems or interfaces such as a call center system. This is a natural impulse, and when envisioning your organization as an AI-first organization, the only limit should be your team's imagination. In the context of MVPs, your number one priority must be to remove as many possible complications as possible to validate the results of the predictions as quickly and cost-effectively as possible. The complexities when constructing these types of processes add up just as easily during software product development as during the initial period when executives need to decide which predictions to select. It is always tempting to try too much too quickly.

Just like the two MVPs already described, engineers additionally should strive to create a Minimum Viable *Process* to move the model results to their end users as quickly and with the least number of complications as possible. One can always add additional integration points to the most minimally viable process after it has been established or replace it entirely with something better. In the case of this customer, we chose to create a simple web app to present the results to the call center agents instead of the equally simple spreadsheet we originally planned. Within weeks, the customer went from not being able to predict anything despite all its rich resources including vast data streams, a skilled data science team, ample budget, and a robust cloud infrastructure to being able to advise when to call any given customer. Now that the AI journey had started, it could be built upon based on the learnings gleaned from measuring the results of the MVPs.

The main goal of the AI-first methodology is to build your ultimate vision of how your organization—enhanced and automated with AIs—can operate in the future. It is your desired to-be state, even if it's not clear how to make it easily achievable if at all. It is your to-be infrastructure and to-be architecture full of different kinds of AIs. The possibilities are vast and wrought with complexities.

On the other extreme are the MVP principles we outlined above. Their one and only purpose is to ensure initial success right out of the gate. They are the first, simplest step toward your AI-first vision. Without them, it will remain a dream and could turn into a nightmare because you cannot make it actionable. It is seductive to spend an infinite amount of time on false creativity, aimless brainstorming, talking, and dreaming of all the possibilities. This is the trap of dreaming. MVPs provide the snap back to reality that allows you to speed toward successful results so that you can build on those very practical learnings. The MVPs we described with this customer example are the perfect tool to fail fast, fail often until you establish early success and bring your vision of AI-first to life.

AI-Ready

The rapid pace of new technologies makes rigid, old-fashioned approaches such as a linear waterfall development methodology uncompetitive for any organization to use for estimating efforts and planning the work ahead. These approaches stifle innovation. The software development community has been trying to solve this age-old problem for a long time. Many methodologies have been invented, tried, and tested with mixed results. The common link between them is an attempt to find a balance between predictability and creativity. If you plan too far ahead, the future you imagined may never appear, and you risk limiting the creativity of your team by too strictly having to follow rigid and prescriptive requirements. Creativity needs flexibility to emerge, yet budgets require it to be balanced with a level of discipline and predictability. If you don't plan and go fully agile, then your team will miss the direction needed to progress, risking the ultimate vision and deliverables. The risk of getting stuck is high, not just somewhere in the middle but in the perpetual beginning of your effort. On average, humans tend to underestimate the complexity of almost everything.

We regularly encounter well-meaning executives who want to achieve this balance. They are often hampered in their ambition to execute on new innovative products or services due to internal bureaucracy, lack of vision, and leadership or political impediments. Trying to be as modern and innovative as possible, they have abandoned waterfall-like approaches, streamlined hierarchies, and provided "innovation budgets." However, progress and speed are still not at the optimal level. The most courageous avoid internal challenges by spinning off their initiatives into quasi-independent businesses. This allows them to succeed as fast as possible without having to watch their backs—or left and right—to appease other departments. Others build "innovation labs"—an internal semi-independent department with relative freedom to apply a typically limited budget for technical experiments. These approaches are valiant as they give a team the focus needed to advance. The third approach we sometimes observe is to add new responsibilities to a selected employee without removing their old responsibilities and duties. The failure rates in these examples we encounter outnumber the success rates by several factors. It is hard to cultivate an idea in such an environment because innovations only happen with a dedicated focus.

Regardless of the approach, progress often slows to a crawl at the line of demarcation between the dedicated innovation teams and the other "mainstream" teams inside of the organization. The latter will have to harden the resulting Minimum Viable Product to be ready for production deployment. As the innovative product moves from the innovation labs into the mainstream department, a string of new stakeholders, who had not been in context from the beginning, get to weigh in. "Just another feature" now needs to be added before proceeding. "Just another review" needs to be conducted. "Just another approval" needs to be attained.

The constant loops of updates and explanations to yet another stakeholder are hard to break, even for the most seasoned of executives. Chief innovation officers without true authority over production environments, or their previous incarnation—chief strategy officers—without control of strategy execution are like R & D without the D. They keep playing with the tools inside of sandboxes. It only becomes innovation if it happens in production.

Risk-averse managers and processes cannot be mandated to embrace risk. There is a valid and important reason for balancing and managing risk when you have a large amount of cash to collect as in this case study. Innovation success is also not just a question of budget. Lavish budgets—and in-office games like foosball tables—are not the catalysts for successful innovations. It is the equivalent of the trap of dreaming at the MVP phase. Paralysis by too much analysis will derail your production delivery. It is remarkable that this feeling is often shared by all without a good solution to it. Instead, one should create the necessary freedom within understandable constraints, not another expensive slideshow from another strategy consulting firm. You have no choice but to innovate from within.

Some of these decelerating factors were present within this financial giant. It would have been surprising if they weren't. The simple tool we deploy at first to ensure AI readiness is a rapid assessment. It timeboxes us and commits every stakeholder—from CEO to data scientist—to focus for no more than a few weeks. Everyone participates in and commits to a mutual discovery of up to three artifacts that will turn your vision into reality, and break through the paralysis by infinite analysis. The risk of participating in such an AI readiness assessment is low. Worst case, you only wasted a little bit of time and a fraction of your budget.

These few, but mandatory, and intensely focused meetings in a discovery allow you to translate your ultimate AI-first vision into a comprehensive outline of your ideal to-be architecture, a clear understanding and documentation of your technological as-is state, and an executable technical road map to get you from point A to AI. The key to accomplishing so much in such a short period of time is to identify the edge cases that bookmark the entire spectrum of possibilities. Instead of attempting to analyze all countries, we selected the two that were at the extremes. Instead of attempting to analyze all communication channels, we selected one. Instead of locking ourselves into all aspects of the customer's infrastructure, we focused on the raw data. Any innovation today is a commodity tomorrow. You must be able to make it happen today.

During these intense few weeks, the customer and our team members investigated the main factors that determine an organization's AI readiness. Accurate and timely data in the right quantity is the main ingredient to feed an AI model. That's why a large portion of the vision depended on the current state, adaptability and flexibility of the client's data silos, IT and business processes, and data pipelines. To avoid being generic in our recommendations, we dug deeply into the unique nature of their business, business model, and end customers. It is always important to marry these joint findings with a good understanding of what their teams had already tried themselves. We did not want to recreate the wheel while ensuring that their team enthusiastically came along for the journey. In parallel, we had to understand the vision of the CEO, and help shape it with the expertise we gained over the years. The suitability of the existing infrastructure to handle the potential new demands from the technical capabilities the AI-first vision required helped us identify the gaps between to-be and as-is, and also informed our precise technical road map.

As the AI readiness assessment neared the finish line, the first set of its "as-is" artifacts was a list of pain points identified and understood by all stakeholders. The current landscape of the customer's infrastructure, existing data pipelines, and DataOps practices—and their imperfections—were documented and clarified. We now had a clear view and understanding of the underlying raw data, its potential value, volume, velocity, and quality. The data's quality differed across countries, and therefore allowed us to home in on the edge cases. The overall high-level architecture was the first artifact our joint efforts created.

The second set of artifacts that emerged from the AI readiness assessment constituted the desired "to-be" state. The high-level architecture diagram of the ultimate vision outlined the dream of what the client's ideal future would look like. The ideal future was not immediately achievable, but what was possible was to derive a clear list of tasks, subtasks, goals, and subgoals from it. All involved understood what problems would be solved when bringing the vision to life. Progress toward the vision would already address pain points iteratively.

The third set of artifacts was encompassed in a road map. The road map had theoretical dates and milestones to implement the ultimate vision, and very well-defined practical steps and estimates to build a quick win—the Minimal Viable Prediction and Minimal Viable Process in our case. Identifying the quick win is where the effort of finding balance should happen. It is not necessarily the easiest task, but rather an achievable one within reasonable constraints. Companies with unlimited budgets and unlimited time often have a hard time to move forward, because it is hard for them to strike this balance of identifying such a small, impactful, foundational first step in line with their AI vision in a short period of time.

With the help of such an AI readiness assessment, it becomes possible to break through roadblocks and direct the creativity of an organization to align behind an understandable vision and bring it to life incrementally without anxiety. The rapid pace of technological development presents an unprecedented opportunity for executives to evolve their organization and business model with a predictable investment and thereby become future-proof.

AI Lighthouse

There is no real golden standard approach to technical innovation. A large, diverse organization embraces new technologies because business environments, technological advances, customer sentiment, and the political landscape are constantly evolving. Each of these creates new blind spots for leaders to prepare for and navigate. Very few are able to start from scratch, having already made large investments into DataOps teams, tools, practices, policies, data pipelines, and large varieties of proper data visualizations. Everything seems to be ready to sail through any storm. But when confronted with a previously unknown modern challenge like the emergence of AI, or other shocks like regulations, competition, or COVID, everything can change, and this seemingly balanced setup exhibits fragility and exposes the blind spots. The perfect order that was perfectly predictable resulted in organizational complexity. It cannot successfully deal with the improbable event, let alone allow the organization to take advantage of it. The entire ship starts tilting, and leaders need to rebalance it fast.

Most leaders we encounter are not oblivious to this danger. They are inspired by these unknown opportunities, yet their organization often freezes when faced with them. They know they need to have the built-in flexibility to make their years of DataOps investments be at their best when it counts the most. The approach that has worked best for us is to create a small, nimble "SWAT Team," constrain it reasonably, and focus it on execution by conducting an intense AI readiness assessment. Because each case is unique, it requires its own custom approach. We make every effort to assess the shortest path to an initial success to build from, reducing complexity and measuring risk. In the case of AI, this assessment helps a leader infuse their vision for their organization with an AI-first mindset that initiates practical first steps rather than attempting to "boil the ocean" by automating everything. Successful leaders avoid being too vast and unspecific in their vision. They start with the creation of a concrete "AI Lighthouse" that guides the way for the rest of their crew to sail forward on their own.

The "SWAT Team" approach cuts through layers of complexity and anchors a leader's vision by creating the organization's Minimal Viable Prediction, Minimal Viable Process, and Minimal Viable Product to show tangible results everyone else can see and replicate. Such a rising AI tide lifts all boats. Organizations that make themselves AI-ready create a flexible system that prepares them to not only insert AI into their daily practices but also gain the ability to benefit from other previously unknown challenges that hide below the horizon.

Executive Summary

Imagine you are the CEO of a Forbes Global 2000 company. You only have time to read three to four sentences. You have all the pieces in place to use AI in your company, including DataOps, different teams, and a modern tech stack, but the results are dissatisfactory. What are the unique tips and tricks from the book chapter that can be a solution for you?

To improve your AI outcomes, focus on adopting an AI-first mindset, prioritizing AI integration across all business processes from the outset. Implement disciplined DataOps practices to ensure agile and efficient data management, emphasizing the importance of extracting insights rapidly. Leverage Minimal Viable

Predictions (MVPs) to quickly demonstrate the value of AI with minimal investment, and conduct an AI readiness assessment to identify and address gaps in your organization's AI capabilities. These steps will accelerate AI adoption and enhance your organization's competitive edge.

Unstructured Data for AI

Case Study Four: Innovative AI Patent Clerk

Imagine your company was sued for patent infringement and is now facing a multimillion-dollar payout to the plaintiff. The question looming before your team is whether to throw in the towel and pay up—or appeal to a higher court.

The five largest patent infringement damage awards totaled $2 billion in 2022, $3.3 billion in 2021, and $4.4 billion in 2020. The infamous smartphone patent wars started around 2009 and continue one and a half decades later, covering dozens of cases involving eleven technology behemoths suing and countersuing each other, and they are only a small portion of increasingly larger patent wars between software and technology corporations due to the rapid pace of innovation.

At this stage one needs to be prepared to successfully appeal and face a panel of three federal judges. It's a huge decision, not only because of the dollars involved but because of the people involved. There is a tremendous amount of time and resources that go into a patent appeal process—and that has a direct impact on the people within the company.

So, you summon your go-to legal team and outline the dilemma, hoping their legal expertise points you in the right direction. "Do we have a strong enough case to bring to the federal court of appeals? Or should we accept the current ruling and cut our losses?"

It's unlikely the team will immediately have an informed answer. It will take weeks to sort through the details of the patent case, your argument, the plaintiff's argument, which law firm the plaintiff is using, which judges will be on the panel, and numerous other details. And after this analysis, they will present you with their qualified legal opinion on the probability of winning the appeal—with an asterisk.

But what if you could create a program to assess those details faster than the legal team could do on their own—and even find details they might miss? What if AI could help you save hundreds of millions of dollars in deliberation by showing you how to pivot the argument in your favor? What, in short, if AI could predict the outcome of a legal case?

Given the power of predictive analytics, perhaps an AI-powered system could improve your chances in court and tell you when to sue, settle, accept a ruling, walk away, or how to adjust the members of your legal team and strategy for better chances of victory? Because of the unstructured nature of such data—court decisions captured in legal documents—applying AI to it requires extensive preparatory efforts. Inherently, for AI engines to predict outcomes based on data, they need to see some kind of structure in the data to identify patterns. There is no structure in chaos, which is what untreated court documents appear to a machine at first. Therefore, after assembling the raw data, every initial phase of AI efforts must focus on predicting outcomes from unstructured data and crucially needs to identify some sort of structure within the data, so the AI can set out trying to detect patterns.

Our client, an AI Patent Clerk, is helping companies predict legal outcomes through a revolutionary platform. Our work with this client allows us to shed light on the intersection of AI and data transformation within the process of Digital Transformation. Specifically, how:

- To process unstructured data
- Time-consuming and important cleaning data is
- To address biases in the data

Gut Feelings

The founder of this company is a Harvard Law School graduate who worked as general counsel for publicly traded technology companies. He was party to potential patent infringement lawsuits and related high-stakes legal proceedings. Over time, he realized that when looking at a case he could often anticipate the outcome based on his experience—gut feelings—by looking not only at the details of the patents but also at the broader scope of who was involved in the case, and many other factors.

Lawyers do not typically practice law based on gut feelings; they build a case off established case law, but experience can point them in the right direction! When cases are repeatable, their outcomes are predictable. The unique nature of patent infringements, where the similarity of cases is not very obvious, makes predictions very hard. To prepare for litigation is very hard, tedious, long work. Attorneys need to find and discern as many facts as "humanly" possible to build their strategy. There will always be unknown factors while creating your own case strategy, such as your opponents' strategy to counter yours. Patent attorneys can only guess what these factors might be and use them as inputs for AI to verify their gut feelings about their strategy.

Although our client's educational and professional pedigree helped him achieve success more often than not, he wondered whether he could automate his gut feelings and use ML to improve his probability of success.

He realized he had an opportunity to help executives of enterprise companies make more informed decisions about whether to accept or appeal a verdict, which could save the company a lot of money and time. Most executives make decisions daily without constantly validating them, but when the decision is as consequential as a federal patent litigation appeal, one has to remove bias and enrich decision-making with as many factual input factors as possible.

You should lean on the best legal advice you can, but even the best experts have blind spots. In addition, their informed opinions could be influenced by current events, gossip, or personal opinions about the lawyers or judges involved. Many factors create bias and negatively influence the probability of accurate legal predictions. Bias prevents accuracy.

What if you could build an AI that supports executives and their legal teams to improve the odds of making the right decision? It would give you a gut check of your gut feelings! As AI is the confluence of data and compute power, the biggest hurdle to overcome will be to find, assemble, clean, and process the data to build such an AI. If successful, its predictions would support these crucial decisions. For every new case, without much budget or time, your C-Suite would make a more informed decision of whether to appeal or not.

Commingled Garbage

Data at the source is typically like commingled garbage. Valuable recycling materials are mixed in with the trash. Extracting the valuables as early as possible in the workflow prevents the need to build much larger "data sorting facilities" further down the value chain, where hundreds of data streams converge.

The challenge is to organize the data extraction sufficiently to determine what intelligence can be derived when overlaying AI. There are a few approaches to this goal based on the volume of data, the amount of work it takes to "clean it," and how much human capital is applied to the task to work within a financial and time constraint. In particular, for extremely large quantities of data, such as the entire catalogue of case law we could work with in this scenario, an approach that applies automated tasks is preferred. The challenge, however, is that there is no certain outcome that any effort will in fact yield recurring and acceptable results.

Many large software and service companies have created entire businesses that maintain these massive, time-consuming, and costly data sorting facilities instead of focusing on the most effective method of nimbly generating data that can easily be transformed into valuable business insights. Of course, sometimes there is no such option. Historical data is often collected in the format of the day, and you have to accept it. This was the case here.

Extracting insights across an array of historical data stored in a less-than-ideal format is daunting and time-consuming. It is tempting to abandon the effort entirely, because you feel the data is garbage after months of effort and no significant gains. This happened to us early on with this client. The US court system is complex and unpredictable. However, if you patiently dig deep enough, you may still find valuable correlations. Our client's convictions drove us to keep digging.

When this client first approached us, the goal was to look at all aspects of the law, from patent lawsuits, to divorces, to traffic tickets. But when you think about all the details and factors at play across all different legal disciplines, the required data set that would need to be sourced, extracted, and transformed into machine-ready format is simply too diverse and vast. It cannot quickly be turned into usable insights.

We decided to narrow the scope by focusing on patent cases—specifically patent *appeals* cases. Only one court in the United States decides these cases, opening up a unique opportunity to create one clear Minimal Viable Prediction—to appeal or not to appeal. Due to the large number of ongoing and historical cases, the data was voluminous enough to possibly contain the insights we were hoping an AI could uncover but concentrated enough with only one data source.

We used this one federal court's publicly available patent appeals cases as a data source and fed clean and clear data into a machine, taught it to understand the various factors and variables at play, and identified patterns to assess the probability of losing or winning an appeal.

The relevant data was hidden in the text. Lots of text. The case documents were tens of thousands of very long PDF documents spelling out every decision of the court—and while there was a certain structure to them, most of them included thousands of pages of unstructured text. Additionally, the format of each document was regularly adjusted over time, which forced the extraction process to be more flexible and sophisticated. Every few years or decades, there was almost always an almost imperceptible change to the document structure, and the process had to be geared toward taking that into account reliably every time.

On top of it, the data itself was extremely nuanced. Countless factors came together in the predictions including the specifics of the case, the relevant parties and their possible histories before the court, and the various law firms collaborating on behalf of each party. Even the judges were considered *as* each panel of judges would interact distinctly against different sets of lawyers from various law firms on either side of the aisle. There are nine judges on the federal court of appeals for patent appeals. Each case has three judges assigned—a lead judge and two supporting judges. Even if the same three judges are assigned in different patent appeals cases, the lead judge might be different, which also impacts the outcome of the case, and therefore were considered when building the prediction model.

In other words, there are a lot of moving pieces. All of this data was applied in the system we designed to draw conclusions and predictions about the possible future decisions and their probabilities.

In some cases, of course, adjusting the variables would make no difference. No matter what, some cases are not winnable, no matter how you shuffle the chess pieces. Even this is an extremely valuable outcome to know a priori. At least the company would then know with a very high probability that they will save millions of dollars in fees by accepting the previous court's decision and avoiding the patent appeals process. As you may imagine, discovering those patterns reliably and predictably is an enormous challenge.

Despite this, it was clear that everything we needed was contained in the case files—who was involved in the litigation, who was representing the parties, the nature of the case, the judges on the panel, each judge's opinion, the arguments, pros and cons of deciding one way or the other, why the decision was made, dates, locations, and long-term impact on comparable cases in the future. Yet until we invented the equivalent of a recycling plant that extracted the valuable parts from the documents, they continued to be commingled garbage without any utility for the sophisticated outcome we wanted.

Structure from Unstructure

Clean data is whatever you can use to build a model on top of, which requires a high level of respect for the data and its *readiness*. Extracting data from its source, cleaning and labeling it correctly, is time-consuming and often the most expensive part of an AI endeavor. And probably the most annoying task in the AI industry for truly creative engineers. Cleaning the data always involves a human; oftentimes lots of humans! Several tens of thousands of people continue to perform these tasks for companies you probably associate with the "automatic" magic of AI, such as OpenAI, Google, xAI, Facebook, Tesla, DeepMind, Microsoft, and countless others. There is no magic. It is hard manual work for millions of people.

By taking an automated approach, we both hoped to accelerate the conclusion on what value we could derive from the data but recognized the very real risk of failure given the complexity and size of the data set. Knowing that it is always hard work to create a Minimally Viable Prediction, especially when dealing with unstructured data, we were not deterred and decided to follow a certain cadence of steps to success. First, we minimized the scope as much as possible before expanding it. Instead of all possible cases, we focused only on the prior ten years of patent appeals. Second, instead of extracting all possible variables before training the model, our plan was to extract the bare minimum, train the model, and validate the prediction accuracy by running the model against historical data. And repeat this routine, if needed.

As feared, our first attempt failed. It turned out that parsing PDFs was even more difficult than expected. There was no one-code-fits-all-PDF because the format of court hearing documents changed subtly over time. We parsed the tens of thousands of PDF files in a semi-automated way and resolved resulting inconsistencies as quickly as possible. Inevitably, our parsers became huge and full of exceptions.

Although the model didn't work for a long time, we were undeterred. We continued to expand the data set weekly until our model reached several thousand parameters that we extracted from *each* case. At some point we were able to iterate and expand to a sufficient point where the outcomes began to show a consistency and logic to them. We achieved this by focusing through iteration on expanding the included variables to achieve an impact as well as enriching our original data set with external data sources like public profile data or patent data.

After every training run of the model, we compared its prediction accuracy to a random prediction. In other words, if you flip a coin to predict a winner, is the model predicting with more accuracy than the coin flip? If it is better, then there is value in the model. If it predicts as well as the coin, there is no value. You may as well save your money and make your decision on the coin flip.

With AI, you're always dealing with historical data, but the power of prediction determines the extent of how far back you need to look to be able to look into the future. With the transit agency example described in chapter 2, it was important for us to have a year or two of historical data, so it could generate accurate predictions, accounting for traffic around annual holidays. But beyond that, going back five or six years, the patterns would not improve the prediction due to variables like road construction or pandemics.

With court data, it was a different dynamic. Patent laws also change, but not as fast as traffic patterns. Instead of discarding it, the oldest available historical data proved valuable to explore. It is important, however, to balance its value with its possible distraction to the training model. We dealt with many years of court decisions and through experimentation and iteration worked our way toward a promising result.

Just as not all historical data is always useful for model training, not all current data is always useful either. It can be irrelevant and end up distracting the model. It has to be removed. A court stenographer certainly has an important role but is an irrelevant variable to extract from a model's perspective.

If a new environmental variable appears in the data, such as a change in the law, and there is no data to represent its implications yet, it is impossible to predict how this new variable may affect prediction outcomes, if any. If you cannot wait for the new law to show its effect in enough historical data to adjust the prediction model, then data scientists, ML engineers, and subject matter experts are indispensable to assess the variable, maybe simulate its impact, create artificial data, train the model to account for the changed variables, and verify that prediction outcomes continue to be accurate.

Ultimately, through hard work and determination, our prediction model far surpassed the coin flip. While we underestimated the manual and semi-manual effort required to get the data ready for training the model by a factor of ten, we appreciated the variety of variables that could lead to an outcome and introduced them in cadence to identify which ultimately made a difference. The new variables that emerged in the data over the years of court decisions required more time from the data scientists and subject matter experts to properly incorporate into the model. Some changes in the structure turned out to be so dramatic that it felt like having to extract the thousands of variables from those cases as if from scratch. All these efforts informed the team to define the successful set of variables that exposed the underlying structure of the data, making it ready for the AI.

Biases

Within the legal system, there is the risk of personal bias that allows for an argument on the interpretation of the law. These differences of opinion are settled by references to case law and the ultimate decision from an arbiter, be it a jury, judge, or panel of judges. In this effort, we were conscious not to build an AI Judge or an AI Attorney but an unbiased AI Clerk to assist human judges and attorneys.

One can easily fall into the rabbit hole of debating the ethics and morals of applying AI to systems, but it is not a focus of this book or this chapter. The purpose of our entire theme *AI driven* and the dissection of this chapter's case study is to make the technical side of AI-driven Digital Transformations approachable to our reader. And to help one make informed decisions about when and where to implement them within an organization. Our hope is to help our readers become AI leaders who can better assess their teams' progress.

Due to the nature of patent appeals litigation, this case study does not contain a heavy emphasis on ethical or moral AI. Rather, the wider context of the legal realm provides good reason to appreciate the importance of bias from a technical and software development perspective. An AI system is not inherently biased. Just like Lady Justice, AIs are blind. When looking at it from an AI system's perspective, it interprets the data it receives to make a prediction with its mathematical models. The bias is in our heads.

The data—be it sales numbers, demographics of a target user group, or the cost of ownership—is being visualized by our brain. This is the moment when the data gets its true meaning, the interpretation of the numbers, words, or images. At this point the reader can subconsciously introduce bias into what is unbiased data.

If this data is then processed by AI, or used as training data for the models, and ultimately turned into predictions—next year's revenue, a sentence to answer a question, or a generated image—the results inherit our biases. Humans interpret the results and may detect or infer bias, but the AI did not add its own bias to the resulting predictions; it has none. It may have inherited human biases due to biases in the training data, or the lack of complete data or overrepresentations in the data. These mathematical biases in the AI's predictions emerge from our source data.

AI is only trained on the provided data; it is not self-aware. We feed the data points into the AI machine and only then does it convert it to predictions. None of the context of our world is taken into account unless we explicitly convert it to some data as well—if even possible—and also feed it into the AI. The AI only processes the input data that *we* put into it. It doesn't know anything about the environment we live in. An AI model is a model, not the actual world we live in.

AI systems project the future based on the slice of the information of the real world at a specific point in time, where for humans-of-the-past the numbers meant one thing, yet for humans-of-the-now they mean something different. The reason is that everything around us is changing constantly. The "unbiased" results of today may scream extreme bias tomorrow and be absolutely unacceptable to humans-of-the-future. The AI is innocent and can't be judged for being biased.

Also, one must always keep in mind that AIs predict future outcomes with a degree of *likelihood* versus the *actual* outcome, which could turn out differently. The purpose of any AI is to make us more intelligent, more informed, and more productive by having a wider array of options with a better understanding of their probabilities. In this area of patent law, we either do our homework manually and get trapped by our unconscious biases or use the unbiased AI tool to gain perspective sooner and build a stronger strategy with a higher probability of success.

Good AI should enhance a human, not replace the human with an AI Judge or an AI Attorney. It should allow us to act on the information it has uncovered, and automate manual processes that are consuming our time, energy, and budget, so we can be more creative and effective. We all have unconscious biases—it's a human trait no matter how fair we think we are. So, if we can direct AI toward our known blind spots first, we can use it to create less biased systems. In a way, this could be a method to gradually uncover and remove unconscious biases from our world, because they get surfaced by an unbiased mechanism, whose "bias" humans can judge.

Starting from Scratch

Over the years, we have interviewed hundreds of highly educated, creative developers, and encountered their very conscious biases toward legacy systems versus their preference of developing from scratch. Without fail, the question of whether they need to support "boring" legacy systems comes up. Not all developers enjoy reading the old, "filthy" code containing the logic of others. They want a blank canvas to apply their creativity.

The reality is that, because of inevitable constraints, blank canvases are just as much a myth as that building from scratch is easier than enhancing a system. It is the hardest part for businesses because you have nothing to compare your progress to. You are comparing the results of your efforts to zero and therefore cannot judge, be judged, or be proud. Only rapid future iterations allow measurement of progress, both technically and with business KPIs, and are therefore easier to assess. Refinement is easier than building into the unknown. The first step is always the hardest.

As discussed in chapter 2, architectures should be future-proof. When great architects build a system from scratch, the temptation is almost too great not to pick the latest and greatest technology stack, the most scalable, the most modern, the most lightweight, the most adaptable, the most customizable, and the most sophisticated. Everything everywhere all at once! In these cases, businesses find themselves held hostage in the future by brilliant, well-meaning, architectural choices that inadvertently lead to job security for the initial team without giving businesses the ability to create future jobs for others, who would have required simpler choices to support you. We have often encountered systems that had to be supported by hundreds of engineers instead of a few, because of non-future-proof choices made when starting from scratch.

Future-proof systems take into account the skills and experiences not only of the first, handpicked, elite team members but also those that come after them—after version 1, version 2, and version n. The skills and experiences of the starter team may be exceedingly rare to find in future generations of developers. There is a reason why you picked them to lay the initial tracks of a new innovation journey. They are presumably the best of the best. We call them "Digital Marines" as they create the beachhead for the rest of the "Digital Military" to follow.

The initial architecture choices have to keep in mind the composition of the team. Software development teams are often composed of retrogrades and innovators. The former prefer to work with previous versions of software rather than the very latest one, because they have been proven. The latter prefer to use the not-yet-released private preview version, because they contain the latest and the greatest features. Neither is right or wrong. It is an art to keep the balance between the two from day one.

Both perspectives have to be balanced, so the system has a low risk of implosion when future generations of engineers take over. Otherwise, they will find it hard to unravel the Digital Marines' train of thought that led to the system they now "own." They will not be able to appreciate its sophistication, recreate decisions that had been made, and fail to get an understanding of the forest for the trees, so to speak. Running a system architected on the cutting edge without weighing the technology choices against the future talent availability makes it impossible to maintain it at a reasonable cost. This is easily demanded in theory, but hard to accomplish in the real world. COBOL engineers (are there still any?) probably demand the highest salaries in the industry for this exact reason.

The key is to reasonably constrain the team members on some dimensions, most notably speed and budget, which vary depending on the complexity of the innovation. In this case study, we not only accepted but embraced the fact that our code would be refactored. We sacrificed perfection by utilizing PaaS models instead of creating bespoke models and utilized a range of preexisting AI services from one of the main cloud providers. We intentionally stayed with one cloud instead of a best-of-breed approach with several clouds to ensure speed-to-market and proof-of-intelligence. CPU power is cheap and continues to drop in price; developers' time is expensive and increasingly so. We chose one of the most common development platforms, .Net, not because we were fanatic about it but because our team knew it like the back of their hands without any ramp-up time that would have been wasted if the team had chosen a faster platform like Rust or Go, or if we had decided to reshuffle the team composition at the expense of speed-to-production. As we have said, our philosophy is that it is only innovation when it is in production. The shortest path to deployment should always win in regard to innovation efforts to avoid the trap of dreaming. The goal should be to prove or disprove the concept as quickly as possible.

Legacy systems that have emerged from future-proof initial decisions on a blank canvas can last decades through constant iterations. They can maintain their elegance and technical beauty with ease of enhancements and maintainability. New engineering team members will be proud to be assigned to them. If there is nothing to improve technically, something is wrong. You will have allowed perfection to get the better of you. We are again confronted with a need to balance between too high of a degree of perfection against the fear of technical debt. Allowing some technical debt is evidence that you have been advancing fast technically while maintaining modernization opportunities for your engineers. Too perfect a system is just as much a yellow flag for experienced innovators as too heavy a load of technical debt that reduces your speed to a crawl.

It may be a bias on our part, but many examples have led us to prescribe the following formula to executives and architects when starting a system from scratch while ensuring it endures the test of time. Choose the technologies that you are most familiar with and are already broadly used as part of your initial architecture. There are enough unknowns to tackle hiding in the details of the problem, so it is not advisable to add technology to the list of complications. Pick team members you know and trust, who share your philosophy and technology choices. From these team members' characteristics, assemble the community of technologists who will support your stack in the future.

With these guidelines, trust your gut feelings, because they will remain timeless without any translation issues. It will be easy for future developers to hitch their own wagons to the train of thoughts that has led from the initial greenfield development to what will then be considered a legacy system.

Reverse Engineering

Brand-name investors as well as top corporations, hedge funds, law firms, and insurance companies provided early capital and profitable revenue streams to the AI Legal Clerk assisting corporate patent attorneys. Its predictive power and clear ROI make a compelling business case. With minimal marketing, word of this disruptive technology has spread to the who's who of American industry, who want to license the software.

This success happened because we dramatically narrowed the initial scope by limiting the number of messy data sources to tackle, thereby creating a future-proof architecture in tight iterations. It was possible to innovate in production with the first sets of real customers within the first year. This helped validate further assumptions we necessarily had to make during research, experimentation, and development. The system continues to be refined with real-life inputs, new variables, and use cases. As initially envisioned and architected, it is versatile to potentially expand into other legal fields, yet the initially narrowed-down target of patent appeals litigation is so vast that it continues to provide a target-rich environment for the company to explore for many years.

Accomplishing the automation of the former corporate attorney's experience has led to a future-proof system that will mature gracefully into a legacy system that will be handed from one technical generation to the next without invoking an engineer's bias toward building from scratch. The bright light that illuminates the blind spots eliminates the uncertainty of choosing a winning legal strategy by intelligently assisting those experts, not replacing them.

During the first few months of technical due diligence, it may have seemed tempting to abandon the data cleaning and tagging effort. But the conviction that there must be valuable recyclables hidden in the commingled garbage made us go back for another iteration to dig deeper. In comparison, knowing that many of the AI efforts that have entered the public consciousness have taken years of tireless iterations, these initial phases of data extraction and labeling went by in a flash. Unlike software architecture and engineering estimates, there were no comparisons at the time to estimate how long it would take to extract value from garbage data. We assembled enough data points to eliminate this particular blind spot. Another *unknown* unknown turned into a *known* unknown for our engineers.

The complexity of AI systems like the one discussed here represents the next level of evolution from the previously discussed case studies. The nature of these types of predictions cannot be as easily and flexibly simulated as bus arrivals. The potential significance of new data points and variables takes much more effort to understand and correctly incorporate into the models. Start training the models as soon as you have some data. Do not wait for all data to be ready. There is no "all." New data will continue to be generated, and needs to be verified, labeled, and incorporated into the models. Its effects will be measured and again be iterated upon, showcasing the endless cycle of successful AI implementation.

Just like with your historical data, removing other irrelevant data is essential, so the models do not use it anymore. Irrelevant data does not mean it is bad or good data. It means that it will have zero impact on the model or will confuse the AI prediction results. AI systems project the future based on the slice of data—the shadow of the real world—we give their models to process. Irrelevant data is one of the ways that overzealous AI efforts accidentally allow biases to creep into their results.

It is imperative to feed only clean, relevant, and timely data into your AI to ensure it makes predictions that have a higher probability of becoming reality. We obviously do not—and cannot—feed the AI the entire world (yet?) to make its predictions, but only a shadow of it in the form of data or words. We expect AI to reverse engineer the real world of tomorrow based on this particular slice of data. Hence the attention, care, respect, and critical thinking that you extend to the data for continuous training of your AI have a direct correlation to the quality of its intelligence and bias. It does not take AI for us to urge you to appreciate this causation.

Executive Summary

Put yourself into the shoes of a decision-maker of a large international company. Your strategy depends on analyzing vast amounts of unstructured data from various subject areas. Summarize the chapter in three to four sentences so it applies to my industry.

In leveraging AI to analyze vast amounts of unstructured data for strategic decision-making in a large international company, focus on preparing and structuring the data to enable AI to uncover patterns and insights. Prioritize data cleaning and the identification of biases to ensure the reliability of AI predictions. Consider narrowing your focus to specific, high-impact areas to efficiently utilize AI for predictive analytics, thereby enhancing decision-making processes across various subject areas. Embrace the iterative process of refining AI models and data sets to continuously improve the accuracy and relevance of the insights generated, supporting strategic objectives, and maintaining a competitive edge in your industry.

Generative AI

Case Study Five: Revolutionary AI-Driven Curriculum Developer

———

Imagine you are a newly minted helicopter pilot, taking control of your highly complex and very expensive machine for the very first time. Your years of training better have prepared you because the impact of one mistake has life-or-death consequences. Training for such a high-risk job is literally a mission-critical job in itself. Your trainers have just as much responsibility as you will have. Your training materials, quizzes, and tests must be comprehensive and up to date.

This case study focuses on building and launching a system of how to use AI to generate individualized curricula within one military branch. Until recently, it took the educators eighteen months to create a satisfactory curriculum for their trainees. Think about all the changes that happen in eighteen months, from technology and operations to staffing and military intelligence. They directly influence protocols and procedures the trainees must master, so the new pilot won't crash the multimillion-dollar bird. By the time they could roll their curriculum out, it was already outdated. By using AI to build the curriculum, we reduced the time to create it from eighteen months down to one week! We were also able, in this circumstance, to leverage advances in large language models (LLMs) to the solution.

We can think of no better way to finish our case study examinations than by introducing you to an inspiring application of generative AI for its intended purpose on an enterprise level—to accelerate and automate creative processes:

- Generative automation capabilities
- Speed-to-production improvements
- Turn AI imperfections into beneficial features
- UI's role in AI
- AI as a painkiller

Time Wasters

Every year this branch of the military required more than one hundred officers to sit in a conference room for several weeks to review the new training materials that would be used to create courses for operators of mission-critical military hardware and physical infrastructure. This involved sifting through hundreds of pages of documents and translating these materials into question banks, which served as the foundation for creating or adapting courses.

The cost to create this curriculum was substantial. The first draft alone took many months to get through. Because the participants had other full-time duties, and these were uncommon activities for these specialists, it took them time to adjust and get into this new context to be productive. The process also required several layers of approvals, which extended the duration of the effort. In the modern technology world, this is considered inefficient and an egregious waste of time and money. This entire process took eighteen months.

While aspiring helicopter pilots were waiting for their most "up-to-date" training, they had to read from the same manual, regardless of their individual backgrounds, skills, prior knowledge, and experiences. The students may have been civilian pilots before, they may have attended aviation classes in college, or they may have not had any prior exposure to aviation at all. Regardless of mixed backgrounds, they were presented with the same set of curricula from the question banks and were graded equally, not appreciating the diversity between thousands of individuals. Incorrect answers did not immediately result in more targeted, individualized questions taking their answers into account. The process continued down its inefficient path for endless months yet again. The goal of using AI to personalize the courses was to ensure that despite different starting points, every trainee reached the same level of competence and proficiency without wasting time on already known topics.

With the current pace of technology, many businesses approach obsolescence at an increasing rate; eighteen months is too long to wait to do anything. The helicopter pilots in question could be confronted with situations or technology they have not been trained on. Upon review, the magnitude of wasted time turned out to be over three thousand hours per contributor to create the training course. The budget for the initial weeks of reviews equaled tens of millions of dollars, mind-blowing!

With everything that we have covered in this book, you now know how an AI assists in accelerating, optimizing, and automating all these time-consuming Sisyphean tasks. While previous examples detailed how to improve the journey by deriving predictions from data, in this case there was an opportunity to leverage generative AI to optimize the end product through iteration. In this case, the AI's help not only became a time-saver but also evolved into a highly personalized training experience for each trainee. Every step of the way, starting with more than one hundred officers assembling in a conference room instead of doing their actual day jobs, became more impactful than the previous process could ever be. Wasting time for the military is simply not an option. The ROI of not wasting time in their context equates not only to saving money but to saving lives.

Progress over Perfection

Perfection is the enemy of progress. AI is not perfect because of its probabilistic nature. It accelerates human experts substantially when applied correctly. Progress, speed-to-market, and rapid, iterative deployments into production should always supersede the trap of seeking perfection. The process of reading manuals is very time-consuming, and today's human attention span has shortened significantly due to the speed technology has brought to our lives. The time and effort required from these officers was obviously enormous and ripe for improvement. It's clear that AI is needed to speed up this process.

Content creation is a much harder task than just reading a manual; add the task of creating questions and answers from the manual and you have a recipe for confusion, exhaustion, and imperfection. You now have to develop different question-and-answer combinations that can teach each student on the most individual level. When creating content, introducing variety and creativity are critical components to consider as there are numerous backgrounds and individual learning styles for each trainee. The creative burden is further complicated by the sheer number of military personnel to be trained.

In addition to reading the manuals and creating content, the third challenge this AI implementation addressed was to define a better process to create educational courses within the military's bureaucracy. Adding AI, at any point, is the equivalent of disrupting the entire process. Without a defined process to adapt, automate, and accelerate, adding AI would have had much less impact. Luckily, the military has a documented process for everything. It's one of those surprising situations where a bureaucracy benefited AI adoption. It is a lesson to keep in mind for many other situations, where organizations consider becoming AI driven.

Not all the facets of these three challenges were addressed by AI equally or even at all. The focus had to stay on AI's immediate impact. In the context of assisting the course creators with their creative tasks of producing data points, after ingesting the voluminous manuals, the AI generated question-and-answer sets. Instead of creating question-and-answer units themselves, the creators could now use the new AI engine's suggestions. Of course, due to the AI's probabilistic nature, they were not always perfect. However, because the human experts used this AI as their assistant, they were able to accept, edit, or optimize the suggestions, allowing the courses to be designed faster with consistent improvement and variety.

With the help of AI, the creative work was now done iteratively. Course creators were able to quickly move from one chapter to the next or from one paragraph to the next while enjoying the AI engine's generated question-and-answer data sets. If they liked what was generated, they added it to a "great example question bank." If they didn't like it, they added it to a "bad example question bank," which helped the AI generate better results. If they found AI's suggestions to be repetitive, they asked it to regenerate the data sets. The variety and possibilities for the course creators became limitless! Of course, the AI-generated results did not bypass the human expert. If its data sets were almost correct, the human expert manually fixed them and moved along with speed. The longer the same AI model was used, the more often it generated great data sets, and the less energy and time it took from the course creators to finish the curriculum. The same question banks—with both good and bad results—will be used to train other AI models in the future when new technical tools will be released. As always, great data leads to great AI, now and in the future.

Hallucination

Imperfections are unavoidable, yet sometimes beneficial. If an AI doesn't know something, it doesn't know how to tell you. Instead, it makes up what it thinks you want to hear. This phenomenon is called hallucination. It applies to all generative AI models, from text-to-image AIs and LLMs to text-to-video AIs and AI generators of other media.

No one in this military branch wanted to substitute officers or instructors with AI, because the AI may hallucinate and therefore give incorrect instructions. You end up with text that the AI generated; however, because these models experienced episodes of hallucination, the text could be meaningless. Hallucinations are real, yet sometimes hard to detect unless you are a subject matter expert in the field that the generative AI is asked to cover. It is hard, if not impossible, to control hallucinations when you rely on LLM. The internet is littered with obvious, even dangerous examples.

At the time of this writing, ChatGPT still cannot correctly resolve the following prompt: "Put the arithmetic operators $+$, $-$, $\times$, or $/$ between the numbers 551022 to equal 100." We are sure you immediately saw obvious answers like $5 \times 5 \times 1 \times (0 + 2 + 2)$ or $5 - 5$

+ 102 – 2. LLMs are not designed for this type of problem-solving. Simple algorithms or your own AIs are a better solution, if you trained it for this task. AI is not deterministic like a calculator, but probabilistic, and therefore will always have a margin of error, no matter how close to 100 percent its volumes of training data and constant tweaking will get it. Not just LLMs suffer from this. Any generative, that is, probabilistic, AI exhibits this phenomenon by design.

In our case study, questions and answers didn't always have to be factual. In fact, the opposite was required. Course creators needed only one correct answer, and while the others had to be incorrect, they could look deceivingly correct to the untrained trainee's eye. Here, the probabilistic nature of AI causing hallucinations posed no danger and actually became a feature of the resulting solution. Hallucinations are not bad unless the AI solution to your business problem requires 100 percent accuracy at all times.

Hallucination is not a defect of an LLM. It is possible to make a generative AI more or less creative by changing the model's parameters or weights. Creativity has business value. If the military chose the AI to be less creative, all questions would be dry and the answer easy to guess. If the AI had been allowed to be more creative, the questions would have been pleasant to read, but not connected to the truth at all, therefore netting zero value. Just like it was accomplished in our case study, you will also always play a little with the "creativity" parameter to find the right compromise. Not all hallucinations are bad hallucinations.

The previous chapters focused on the large efforts of how to make AIs' predictions as precise as possible. An AI predicting bus arrivals is not allowed to be creative. It must be as accurate as possible. With LLMs, the quality of its predictions is measured subjectively for subject matter experts. For instance, does the response sound good enough? The response can come across as extremely convincing to a human subject amateur, which is one of the risks of AI. Subject matter experts must be kept in the loop at all times. Due to the need to be creative and generate different types of answers, precision became an enemy in this case. Benchmarks for generative AIs measure the number of data points they have or the speed of the AI's prompt executions. Quality is starting to be measured, but it still subjectively lies in the eye of the beholder.

Thanks to the popularization of LLMs like ChatGPT, "AI experts" were born overnight, and the concept of LLM proved to be useful. Several large corporations released their own LLMs, some of which were open sourced and already caught up with ChatGPT in terms of data points and quality. Some LLMs can even be used for commercial projects. Building LLMs is an expensive multimillion-dollar process, involving thousands of people, billions of documents, and staggering electricity costs. For the majority of businesses, there is no need to build their own LLM from scratch. Licensing and using ready-to-use LLMs from major vendors is not expensive. Limited demo versions even run on your laptop, probably far too slow for what you require in production, but fast enough to prove their usefulness as a possible concept to accelerate and optimize your business processes.

In our case, it was not clear which of the models would provide the highest level of quality—remember, the quality is subjective! From a high vantage point, each model is supposed to do the same thing. They look similar, but each works a little differently. We tried all models available at the time to find which one would provide the highest level of quality. The previous curriculum materials, created by subject matter experts, were used to train the models.

LLMs are different from other AI models, because they work with free-form text data that appears unstructured to us. Under the hood though, there is structure. The *large* amount of text data that is used to train *LLMs* enables them to learn the patterns and *structure* of natural language and infer the relationships between words and context.

For the initial model training, on top of generic LLMs, only one manual including three hundred questions and answers was used. The results were not perfect. While experimenting further, promising results appeared only after we reached one thousand questions! This is an extremely small data set when compared to the equivalent of about one hundred million *novels* of text data that ChatGPT was trained on. Even on such an enterprise level, it is efficient to use your own data on top of existing LLM options. After the initial model training, a new manual was used to make the models generate samples. The instructors validated the subjective quality of each of the model's results to get as close to what a human would have created pre-AI over many months. Some models hallucinated too much, even when playing with "creativity" parameters. Others produced plausible results, even making the AI's outcome indistinguishable from a human instructor's.

With the experimentation phase behind us, and the risks mitigated, we entered the architecture and engineering phase of the effort.

Hallucinations are par for the course when playing with AI, because they are not factual. It is a risk you can mitigate with the right steps, the right features, the right process, and the right people. Hallucinations are not a valid reason to avoid using AI.

Good UI, Great AI

As stated above, before the introduction of AI into the analysis of voluminous manuals, the initial course creation, the fine-tuning, and the final release required an approach to create one course that fit all participants. As a result, it took each trainee and teacher a lot of time and effort to reach their education goals. It resembled an attempt for one teacher to educate one classroom filled with thousands of students all at once. One of the goals of adding AI to this yearslong, recurring exercise was not only to accelerate the course creation process but also to create diversified courses, even to the point to tailor to a classroom of one. The idea of personalization was a new concept to this branch of the military. As a result, each student could interact with this system in the most individualized fashion possible.

The effect of adding this type of *Artificial* Intelligence was not a more artificial experience for the trainees; it was a more *intelligent*, individualized experience. Suddenly, the accelerated AI-assisted teachers were able to create courses that took every student's background into account, resulting in a highly bespoke, authentic learning experience. Individuality instead of uniformity.

Each trainee learned at their own pace, with the right question-and-answer combinations, focusing more on what they didn't yet know than what they already knew. The minute they understood, the system moved them right along. If the trainee didn't answer correctly, the system would present them with a new, tailor-made set of follow-up questions. If they got stuck, with the infinite patience only machines are capable of, the new approach could spoon-feed the trainee with new information to help them advance. Suddenly it felt like the student-to-teacher ratio had gone from 1:1,000 to 1:1 in the blink of an eye.

AI on its own is useless; you have to interact with it. While certain AIs manage traffic to create optimal green lights for commuters, the commuters will not directly interact with this AI or even be consciously aware of its presence and benefits. The traffic planners in a backroom are the lucky ones experiencing the AI-awe.

AI is made and built for humans. Therefore, humans need tools to interact with AI. Our case study had a web interface that the students used to consume their courses, answer questions, and take assessments. They experienced a more personalized journey, learning new skills. You *must* have a UI to *really* use AI.

The other equally important UI is the one between the instructor and their AI assistant. In this interface, it was easy for the instructors to mark correct or incorrect question-and-answer suggestions. This is what's called Reinforcement Learning from Human Feedback (RLHF), an indispensable part of successful LLM-based AI systems. It is the way to ensure that AI learns from the subject matter expert what is "right" versus what is "wrong," what is relevant versus what is not, what are hallucinated facts versus true facts, what are useful hallucinations, and which ones are too far off the mark—and can improve the quality of its output over time. This interaction between the AI assistant and the educators provided the AI an opportunity to receive the subject matter experts' feedback it needed to learn. This is called fine-tuning of the model, which is required to guarantee success.

The AI will generate questions and answers for you but requires input from the expert educator to make them truly impactful and able to build upon. The initial training of the model is the vast majority of the effort to tune the AI to become impactful. This crucial UI-based interaction of fine-tuning and reinforcement learning enables the final polishing of the outcomes. By constraining the LLM and "forcing" it to try harder, its suggestions became more right than wrong more often than not.

With these initial constraining and enforcement steps, the AI-generated courses were released to the trainees without subsequent and ongoing hyper-monitoring. Without these UIs to constrain, polish, and expose the AI, there was no way to create highly personalized courses for each trainee. The instructors' unique skills and experiences scaled from a one-size-fits-all paper-based course to a digital trainee experience that gave every student the feeling of their own personal tutor.

True Engineering

With traditional development, you can use virtually any cloud, and their capabilities and costs are roughly the same. In comparison, LLMs and their generative AI siblings are compute-heavy technologies. They require costly and unique infrastructure layers to reveal their capabilities for your business. They introduce instability in certain circumstances, running too slowly or becoming too expensive for the specific task, reducing your ROI.

The right combination of the various mosaic pieces you must consider and assemble to complete the ideal picture overall depends on your ultimate goal of introducing a specific AI into a specific business process. A different AI and a different process may require a different set of mosaic pieces to assemble. Therefore—as we outlined in chapter 1—your various interlinked architecture layers and constant experimentation hold the key to your success. Solve these complexities with focus, patience, and perseverance and your AI solutions will appear deceptively simple.

While the right architecture holds the key to your AI solutions, your team members' skills, experiences, and composition hold the key to successfully maintaining them. Putting the microscope on our case study, each engineer had a critical role and was introduced at a specific time in the data pipeline. Data scientists oversaw all aspects of an integrated process across the entire pipeline. Data engineers represented the beginning of the data pipeline and were responsible for extracting the most meaningful data from various data sources. ML engineers—sitting in the middle of the pipeline—use data to train models. MLOps engineers are at the very end of these pipelines, deploying the ML engineers' results into production in the most predictable and automated way.

MLOps is very close to DevOps in this regard; some call it the DevOps for ML. MLOps automates development and production deployment of models, and is responsible for their constant monitoring, validation, and governance. The MLOps discipline is one of the youngest of the various *Ops disciplines: DevOps, DevSecOps, DataOps, FinOps, or GitOps. DataOps experts oversee and enforce how to treat your organization's data across the various engineering teams. DataOps is mostly a data governance discipline that formalizes all aspects of data management. MLOps is a true engineering discipline that is responsible for deploying and monitoring models in production.

There are many options where the models can be deployed. Most public cloud vendors provide the necessary hardware options that are optimized for ML execution. Executing models are heavy on parallel compute calculations. As a result, they don't run fast enough on traditional CPUs. The nature of ML calculations is very close to what GPUs were designed for, which is why many models work great on GPUs. GPUs are not the only option. There are also specialized chips available, like TPUs (Tensor Processing Units), IPUs (Intelligence Processing Units), NPUs (Neural Network Processing Units), and other *PUs— all of them were designed for highly parallel processing requirements.

The fundamental puzzle to solve with modular architecture, hardware requirements, and the right team composition is to select the right cloud infrastructure and service platform with cost-efficient hardware for your AI business case. A "regular," "traditional" cross-cloud or On-Premise-to-cloud migration is already a complex and expensive journey. The finesse of a migration with heavy AI models or predictions is even trickier to achieve due to the volume of data required. Fortunately, in our case, our own training data of the manuals and questions was relatively small. We uploaded it to all major clouds for a test run to measure both performance and cost, as well as choose the most applicable cloud for the project. In theory, clouds with specifically AI-optimized chips should be more economically viable. However, in our case, running our model on GPUs turned out to be the cheapest and most stable. We struck the right balance to optimize an acceptable performance for the effort.

There are cheaper options than "renting" hardware from cloud providers. Using public services like ChatGPT, Midjourney, DALL-E, GitHub's Copilot, or others could be a solution. However, every time you submit a new prompt on one of those services, you voluntarily submit your organization's potentially sensitive information to a third-party vendor. In our case, due to the highly sensitive nature of the training materials, this was not an option.

There are a lot of ideal theoretical solutions that may look really good on paper. Sometimes you might be lucky, and your theory holds true in practice. Dreams can come true, and some people do win the lottery. We, too, love doing guesstimates and conduct intellectual exercises. But we also know and love having to verify all our ideas in practice and in production. This means that choosing the right infrastructure and the right team structure along the process we call pipeline has to fit every particular situation. Altogether, this gives you predictability.

As mentioned in chapter 2, our approach is to constantly iterate through all aspects of the unknowns, running small experiments, checking every single step, and turning unknowns into knowns. This is what true engineering on the cutting edge is all about.

Painkiller

Time-consuming tasks such as reading an in-depth manual of how to operate extremely complicated machinery, coming up with wrong answers on purpose, or creating a summary of a chapter were a draining exercise. Many of the participants in this process had to do this outside of their daily, mission-critical responsibilities. Even though the process was repetitive, it required originality and creativity. It could not be transformed into a mass production assembly line. The participants had to master their craft to produce excellent results.

These creative, yet painful, tasks exist in many of the business processes each of us encounter on a daily basis. Entire "capture teams" are dedicated to uncovering, analyzing, and responding to RFIs or RFPs. Many organizations issue or respond to proposals or grant applications. Most employers have a lengthy performance review process that disrupts the day-to-day routine of the participants. In each case, there is always something new to consider by all those involved. Your business is filled with such creative, painful, repetitive, time-consuming, and expensive routines that require bureaucracy. These routines cannot be replaced or avoided, but they *can* be optimized, and the pain *can* be lessened by the assistance of AI.

To begin such a nonroutine process typically presents an uncomfortable, significant mental effort to experts, who otherwise have distinctly different daily tasks. It took the officers time to get into context and "warm up" before they hit peak productivity. The same is true for any of the participants of similar business processes we mentioned above. The way generative AI—in our case based on LLM—alleviates this pain is by assisting the experts to accelerate, optimize, and automate the pain points to unleash the participants' imagination and creativity. It allows the creators to get started more easily without

119

fearing the risk of failure that could block them. Imagination cannot be mass produced, but it can be sparked by intelligent use of AI.

Equally as important to unleashing the initial spark, you also need to stay focused and move the process along efficiently. This is how the UI that constrains the AI aids the military's expert users. It ensures that they don't get overwhelmed, confused, or distracted by too many options to apply their expertise. The simplicity of the UI wins over the complexity of the task. The simple interface allows the expert to easily navigate and focus without additional training or supervision.

AI had such radical effects on accelerating the tasks that we were able to bring a process that took eighteen months down to achieving iterative results that would take only a week to work through, because the process was clearly defined and documented. It gave this AI solution a greater opportunity to reduce an ineffectual organizational burden. Reiterated from above, despite a concern that adding these nonhuman automations, analytics, and Artificial Intelligence to a personal experience might feel artificial, the opposite happened. The thoughtful and systematic removal of pain points led to more intelligence, more individualization, and more satisfaction.

An expert was kept in the AI loop at all times, which allowed the system to prove its ROI in production without risk. Whenever you apply AI, it can change things dramatically and reduce organizational and individual pain caused by creative and time-consuming processes. It would have represented a much higher risk to wait for the perfect system instead of applying AI to accelerate and assist with the obvious tasks immediately. In this case study and many comparable cases, embedding an LLM into an AI system proved to be a real painkiller, not just a fancy supplement.

Executive Summary

I'm a top executive of a well-known global brand. From the chapter above, what are the main characteristics, pitfalls, and benefits of how generative AI can improve my company? Provide the summary in three to four sentences.

Generative AI can significantly enhance your company's efficiency and innovation by automating time-consuming tasks, providing personalized experiences, and rapidly updating content in response to changes in information or protocols. Its flexibility and adaptability make it invaluable in areas requiring creative output or where information is constantly evolving. However, it's important to be mindful of its probabilistic nature, which can lead to inaccuracies or "hallucinations" in generated outputs, necessitating oversight by human experts. Despite these challenges, the benefits of speed, personalization, and the ability to transform cumbersome processes into streamlined operations present a compelling case for its adoption in strategic areas of your business.

CONCLUSION

Using these five case studies, we wanted to take the reader on a journey to understand how AI is impacting industry or even your own company. We introduced best practices, strategies, and methodologies for implementing AI within an organization and how to move into production as fast as professionally possible. They have proven to work for us many times. Our hope is that these real-life examples provide enough inspiration for any leader or participant to take the small leap of faith to start experimenting with AI. The broad selection of customer use cases should make it easy to identify similar opportunities within your own field of expertise.

We have fundamental conviction that whether your journey leads to an AI-driven business or not, Digital Transformation is inescapable. We feel that we are at a comparable point to when Steve Jobs introduced the iPhone on January 9, 2007. Back then, no organization was mobile-ready, let alone mobile-driven. Eventually virtually all jumped on the mobile train, and many ran off the platform to complete the leap. Becoming AI-ready by embarking at long last on a serious Digital Transformation journey is equally urgent; becoming AI driven is imperative. The AI train is in motion, and there is still room to jump aboard by adopting the learnings from this book and recognizing the vast opportunities ahead. If you do, we are convinced that you will celebrate becoming a category killer in your industry.

In walking through the logic of Digital Transformation, AI readiness, and creating an AI-driven business through these examples, you should be able to determine the maturity level of your digital destiny. This book can serve as your guide to understand your immediate next steps, strategies, and approaches and inspire the rest of your organization to become AI driven. The examples we mentioned and the tools we methodically used to tackle the various challenges will allow you to better plan your own road map and accelerate your teams' execution of your strategy. This logic applies to everyone with the determination to put one foot in front of the other.

Chapter 1 outlined step-by-step how to start and conclude laying the foundation to adopt and create modern technologies. It is often called Modernization, Innovation, or Digital Transformation. Thereafter, the subsequent four case studies walk the reader through applied AI on an enterprise level. They focus on the distinction between structured, unstructured data and generative AI, and explain the key methodologies one should apply that allow AI to derive value from data. By reading or rereading them, you can identify which approach best applies to you to start your journey.

Chapter 2 demystified AI through the lens of one of the most understandable approaches to data, namely deriving value from *structured* data. We have encountered a lot of confusion when working with business leaders about what AI is; the case study we picked was both highly relatable to any user and served to highlight how to derive increasing value from structured data through pattern recognition and advanced data analysis. We also used this as an opportunity to introduce key AI terms. Step-by-step, the case study shows how this type of predictive analytics can be expanded by using more advanced ML technologies and methodologies within core business functions, eventually providing comprehensive future business decisions based on highly probable predictions. An organization's flexible system architecture, scalable infrastructure, and various optimizations and automations from using DevOps best practices are key technical prerequisites for such an AI success.

Chapter 3 introduced additional AI and data methodologies and terms that rapidly drive immediate business outcomes, preventing "paralysis by analysis" and future-proofing an organization. The goal of any AI effort should always be to generate a positive ROI, justify the investment of time and resources, and be able to adjust with utmost agility. At first glance, everything looked in order. The customer had control of their data, a proficient Data Science department, and adhered to DataOps practices, yet tangible results were elusive. Breaking the established order was necessary. Only when we helped create a robust shortcut between data scientists and the consumers of their work were the predictions accelerated, validated, and therefore informed further data acquisition and model creation. The customer was "unstuck" and could apply the learnings from these quick wins to adopt similar approaches across the organization.

Chapter 4 details how to derive value from the other side of the data coin—*unstructured* data. To process, clean, and find structure in unstructured data is time-consuming and can be surprisingly expensive. Preparing the data for AI to derive actionable results can be prohibitive without a defined, methodical approach. Garbage data-in leads to garbage predictions and biases out. Every AI model contains structure and logic under the hood; therefore, you need to exert a meaningful, targeted effort to discover some structure within the unstructured data set and apply it to train the model. This prepares the model to discover patterns in the production data you will use for predicting the outcomes of your business case. AI is only as good as the quality of your data.

Generative AI using LLMs is a radically different facet of AI and was introduced in chapter 5. The fundamental difference to the way we outlined in the other approaches is that it is not required to be precise. *LLMs* contain tens of *billions* of variables and therefore use immense compute power and energy, not to mention the rapidly growing workforce needed to optimize these models as they grow. As a business, you can leverage these massive third-party investments for your own purposes instead of creating your own LLMs. It's truly mind-blowing to know that these models are open sourced, available for anyone to leverage. Building an LLM yourself at this stage is unnecessary and likely unachievable in any competitive fashion. "All" you need to focus on is to fine-tune an open source LLM by labeling *your* training data. This allows the LLMs to be smarter as they look for patterns in your production data.

Commit

Evolving technically, while unavoidable, is a challenge, and our hope is that in reading this book you find yourself better prepared to overcome those challenges. As our case studies have guided you through the preparation for change, a clear vision and a step-by-step approach is the best recipe to ensure success at the lowest cost, and we want to stress some elements of that advice.

While adopting AI is imperative for any organization, the risk of implementation failures is most acute early on and in particular within an organization with insufficiently formalized or documented business processes. Poor documentation usually comes with one clear result, that the critical knowledge of your technical environment is retained in the heads of a few key people, which ultimately holds you hostage to their availability. While you can quickly advance on a decision to technically evolve toward AI, your first priority is to concurrently commit to your business vision as a team and culture and establish documented processes to support it and reduce the cost of indecision.

In addition, defining the right team to carry you through this journey is another essential element to contain risk and cost. A flexible and extendable architecture and well-defined processes are not enough to adopt AI. As a best practice, you need to form an AI SWAT Team of your elite internal and external experts. With this "tip of the spear" approach, your best team members chart the way for others to follow while keeping the capabilities of the next generation of engineers in mind. This does not mean that your existing team becomes obsolete; to the contrary, it begins to free up individuals and provides them time to train and upskill in the wake of your elite techies regardless of if they are in-house or contracted in for a period.

At all times, subject matter experts must continue to supervise your AIs, because AIs only work with an interpretation of *our* interpretation of the world as *we* see and monitor it with data. The data only describes a shadow, a small slice of the world. It is as if AI observes the world through a keyhole. At this stage, it can never see the whole picture. People must stay in the loop to avoid errors from creeping in and to maintain control that the predictions stay accurate and meaningful to the users of your AI in the real world and continue to generate a positive ROI for your organization.

While the hype created by public generative AIs using generic data has illustrated the power of predictions to a very large audience, in our opinion the bulk of positive returns for enterprises lies in leveraging their existing data. As we described, there is an approach to both structured and unstructured data, both of which yield tremendous efficiencies and powerful—even surprising—outcomes when harnessed correctly, that can add rocket fuel to your business. The majority of "low-hanging fruit" and more substantial ROI opportunities for large organizations are in using AI to accelerate, automate, and transform their vast array of business processes. Generative AI has its role to play and is worth investigating, and we find ourselves in early explorations of where it can unlock hidden value in a client's data.

Finding approaches to leverage open source LLMs is without a doubt the best path to capitalize on this revolutionary evolution of AI. Importantly, as we described, it's in most cases essential to define and develop a well-designed user interface to generate optimal results with generative AI. Such a UI for AI constrains the users on their essential input and feedback that will enable the LLM to create the most impactful output. It will help them use this type of AI most efficiently and achieve the most effective results. Such a constraining and guiding UI reduces the barrier of starting and completing many of the previously painful tasks, thereby radically optimizing productivity. When applied to accelerate similar high-value tasks, LLMs can boost productivity rates within an enterprise significantly. In our case study, it was literally one hundred times.

Fully committing to transforming the very core of what you do means that you have just invoked one of the most disruptive changes your organization will ever undergo—and it will never come to a halt again. Executing the right AI-driven Digital Transformation will convert your current company into a software product company. Should your organization go far enough to become an AI-driven software company, your investment priorities will need to shift. Technology and R & D budgets of software companies are magnitudes larger than traditional companies' IT budgets. But as we said, the path there should be measured iteratively to ensure wise investment while giving sufficient time to realize the savings and returns from early wins to reinvest into future needs.

Digital Transformation translates into creating a future-proof architecture that allows the creation of services and systems that generate reliable and sufficient data to train your AI. *AI-driven* Digital Transformation focuses on extracting value from your data in the form of actionable information and reliable predictions. Your architectural choices are a fundamental key to your success. It mirrors your organization's structure and vice versa. Business processes that AI automates allow you to achieve your organization's goals by outlining the specific activities and necessary workflows.

Business processes are deterministic by their very nature. The decision nodes within them are probabilistic: Of the three possible decisions you have to make, one is more right than the other, but not always. You have to make as informed a prediction of the outcome of your decision as possible. Yet predictions are just that: predictions. They might work, they might not. If they do, they may only help you make better decisions for a period of time because the world at large is constantly changing. However, relentless recalibration of the AI models inevitably leads to improving and more valuable predictions that can unlock exponential growth.

The bottom line is AI is not magic; it's hard and smart work to reap its benefits. But in the end—when following one or a combination of the five playbooks we described—the results regardless are magical. With all the ingredients for success in place that we have highlighted in this book, it can take months of unrelenting focus and effort to achieve the first promising results. Such a defined and deliberate approach will improve your current technical systems and capabilities while maintaining your current business without interruption. A holistic AI-driven vision as we advocate with this book will allow you to take ownership of your digital destiny in perpetuity.

Ⓡ Evolution

We live in an age where the digital asteroid has struck. Its fireball is spreading at internet speed, devouring one previously unbeatable deterministic T-Rex after another, one prehistoric species after another. However, some not only survive the onslaught but thrive—without resorting to luck, magic, or hope. The evolution of deterministic dinosaurs to probabilistic mammals has begun. In chapters 1, 2, and 3, we investigated three dinosaurs, who evolved successfully and explained how they did it. They now live in the same period as the two probabilistic mammals from chapters 4 and 5. We chose these case studies to encourage you to embrace similar evolution.

Every aspect of our professional and private lives is increasingly influenced by AI directly or indirectly. We probably aren't aware of it. The scale and texture of this influence are so big and nuanced that large swaths of the global economy have already become AI driven, which is why we chose *AI Driven* as the title of this book. It is just a matter of time for the remaining dinosaur organizations and their prehistoric species—a.k.a. industries—to follow or become extinct.

Every so often in the IT industry, the convergence of new capabilities—tools, team members' hard-acquired skills and experiences, new approaches, algorithms, models, cloud offerings, hardware, innovative use cases, process documentation, or widespread respect for and availability of good data—ignites a new evolution with a fundamental, long-term impact.

As outlined throughout this book, AI in production is the most recent long-wave IT cycle that layers itself on top of other long-term

(R)evolutionary IT arcs. One of the earlier trends with long-term impact that you may have witnessed was the ongoing decades of web-driven app development, the two decades of mobile-driven innovations, or the cloud-driven evolution. These long arcs don't replace one another; they multiply each other, creating a compounding aftermath that magnifies their individual impacts.

These evolutionary bursts of innovation along with lasting effects across industries pose an existential risk and outsized opportunity to every single business today. New business models, alternative revenue streams, and entirely new industries are emerging from these digital reincarnations of Schumpeter's gale, the famous eighteenth-century Austrian economist's concept. These innovations foster new waves of creative destruction or rather *disruption* of obsolete innovations with modern technologies.

Throughout this book we endeavored to explain AI, suggesting actionable steps and processes to increase the efficiency of and innovative posture and flexibility for any type of business. Already in his *Origin of Species*, Charles Darwin observed that "extinction of old forms is the almost inevitable consequence of the production of new forms." Our goal was to show legacy companies how to leverage the adaptive radiation of technological disruptions for their own mutation. These internally pushed evolutions can result in external revolutions—new or existing businesses creating new intellectual property they can patent and trademark. Most people think history repeats itself. We think it revolves in virtuous circles.

Continuous experiments in production systems with IT innovations such as AI can prevent your organization from becoming your industry's Sue, the T-Rex, in Chicago's Field Museum. The next unicorn disrupting *your* industry might never happen, but only *if you* are ready to preempt its attack by adopting the tools, frameworks, and approaches we outlined here.

The media company from chapter 1 in slow decline, vulnerable to attack from innovative start-ups, created a substantial new software license revenue stream and thereby turned itself into a software company. By disrupting its very own legacy self and industry, it exemplifies the dinosaur not only surviving the potential cataclysm but using it to its advantage. The transit agency from chapter 2 and the financial company we introduced in chapter 3 show how adopting these sophisticated technologies does not have to be a make-or-break decision but can be a winning formula for success—instead of becoming victims of digital extinction events. Without the power of cloud computing, good data, and the ability to uncover new business value using AI, the start-ups from chapters 4 and 5 would never have been created.

IT as a vertical silo has turned horizontally, penetrating other non-IT vertical silos. Disruptors and innovations ignite where the two silos intersect. AI is one of the forces behind these changes, pushing businesses to evolve and adapt. This thrust into metamorphosis has been accelerated by AI. The next industries to emerge may be robotics with a less over-engineered version of Rosie the robot, decreasing the number of household chores. Or it may come in the health industry in the form of food as medicine, making us all forget the concept of pills. Or it may be the space industry, if space innovations can lead the way to a positive ROI, opening an opportunity for a variety of enterprises to replace the current, subsidized government-sponsored agencies. Only your creativity and the future will tell.

This new version of the digital world offers us the opportunity to take ownership of our own evolutionary destiny as businesses, leaders, and category killers. Dinosaurs, that is, legacy entities, tiny mammals, that is, start-ups, and any other case we covered in *AI Driven* can each survive and thrive in an era of creative disruption. The greatest safety from AI-driven and the next generation of extinction events lies in the speed of evolving the organizations we lead.

The tools at the disposal of any institution that is ready to become AI driven will continue to evolve. Hallucinations in LLMs may be eliminated by further improvements of Retrieval Augmented Generation (RAG) patterns or similar advances. Copyright laws may adopt the influence of generative AI and find ways to treat it respectfully and safely without strangling its progress. The use of AI will certainly be better governed, maybe mirroring how we learned to regulate car safety. Every enterprise will accelerate its transformation to warp speed to continue on its mission, to explore unique new opportunities, to seek out new businesses and establish new industries, to boldly go where no one has gone before!

ACKNOWLEDGMENTS

Spending two years to write *AI Driven* was a transformative journey for us. We had no idea how difficult it would be to distill countless experiences into a comprehensive book. We tried to be careful to only select examples that matter. In fact, we rewrote the book three times. The two of us sacrificed many weekends and holidays with our families and friends to reach our standard of quality. We are proud of the result, and hope you find it valuable. It would not have been possible without the unique talents and insights of a remarkable group of individuals, whom we would like to extend our gratitude to.

Direct contributors are in alphabetical order to underscore that we equally value and appreciate each of them:

- Paolo Capocasa for his artistic illustrations of artificial intelligence and the Digital Black Swan book cover.
- Xenia Gray for the nonmagical, yet conceptually accurate illustration of AI engineering as a multifaceted rabbit.
- Jennifer Iras for shortening mile-long sentences into precise statements (we admit some of them are still long).
- Kiryll Kraikivski for sharing his deep expertise in paleontology and evolutionary science with us.
- Philippe Lanier for volunteering to be our hyper-critical, hyper-focused Über executive test audience.

- Mikkel Pind Schultz for helping throughout the first iteration of writing the book and the creative spark that led to the title *AI Driven*.

Efforts of hundreds of team members over decades of impactful projects have formed our understanding of the power of AI.

Friends and thought leaders, who have inspired us throughout our careers without maybe being aware of it: Michael Afsah-Mohallatee, Stanislav Arseniev, Ken Babby, Abu Bakar, Sid Banerjee, Sanju Bansal, Ryan Bateman, Jeff Bedell, Vincent Bezemer, Julia Beizer, Bret Beresford-Wood, Anthony Beverina, Matt Cascio, Ken Chen, Ernest Chrappah, Aneesh Chopra, Yael Cosset, David Do, Maria Eitel, Kirill Evstratov, Randy Frazier, Alex Freixas, Jason Gabor, Carl Grant III, Alan Harris, Christian Hernandez, Stewart Holness, Darren Jaffrey, Jody Kalmbach, Jonathan Klein, Niki Kollia, Cyrus Krohn, Vivek Kundra, Karl-Heinz Land, Anthony Lanier, Adam Levy, Lang Ly, Igor Lychagov, Andrey Maltsev, Vladimir Menshakov, Frank Møllerop, George Moore, Kimberly Nelson, Andy Norris, Klaus-Anders Nysteen, Shailesh Prakash, Clemens Praendl, Ram Ramachandran, Braxton Robbason, Maksat Rysaliev, Eduardo Sanchez, Michael Saylor, Alex Shkor, Shane Shook, Marcel Smit, Arnold Sowa, Ann Timmons, Cole Van Nice, Mary Kaye Vavasour, Katharine Weymouth, Kelli Wheeler, Mike Z, and dozens of customers and partners.

www.ingramcontent.com/pod-product-compliance
Lightning Source LLC
La Vergne TN
LVHW092005050326
832904LV00018B/323/J